PROBLEM
SOLVING
in
MATHEMATICS

Cover design by Bev and Charles Dana

This work was developed under an ESEA Title IVC grant from the Oregon Department of Education, Office of Policy and Program Development. The content, however, does not necessarily reflect the position or policy of the Oregon Department of Education and no official endorsement of these materials should be inferred.

Distribution for this work was arranged by LINC Resources, Inc.

ISBN 0-86651-186-5

Order Number DS01412

10 11 12 13 14 15-MA-02 01 00

DALE
SEYMOUR
PUBLICATIONS
P.O. BOX 10888
PALO ALTO, CA 94303

PROBLEM SOLVING IN MATHEMATICS

PROJECT STAFF

DIRECTOR:	OSCAR SCHAAF, UNIVERSITY OF OREGON
ASSOCIATE DIRECTOR:	RICHARD BRANNAN, LANE EDUCATION SERVICE DISTRICT

WRITERS:
RICHARD BRANNAN
MARYANN DEBRICK
JUDITH JOHNSON
GLENDA KIMERLING
SCOTT McFADDEN
JILL McKENNEY
OSCAR SCHAAF
MARY ANN TODD

PRODUCTION:
MEREDITH SCHAAF
BARBARA STOEFFLER

EVALUATION:
HENRY DIZNEY
ARTHUR MITTMAN
JAMES ELLIOTT
LESLIE MAYES
ALISTAIR PEACOCK

PROJECT GRADUATE
STUDENTS:
FRANK DEBRICK
MAX GILLETT
KEN JENSEN
PATTY KINCAID
CARTER McCONNELL
TOM STONE

ACKNOWLEDGEMENTS:

TITLE IV-C LIAISON: Ray Talbert
                             Charles Nelson

                             Monitoring Team

                             Charles Barker
                             Ron Clawson
                             Jeri Dickerson
                             Anthony Fernandez
                             Richard Olson
                             Ralph Parrish
                             Fred Rugh
                             Alton Smedstad

ADVISORY COMMITTEE: Mary Grace Kantowski   University of Florida
                             John LeBlanc   Indiana University
                             Richard Lesh   Northwestern University
                             Len Pikaart   Ohio University
                             Kenneth Vos   The College of St. Catherine

A special thanks is due to the many teachers, schools, and districts within
the state of Oregon that have participated in the development and evaluation
of the project materials. A list would be lengthy and certainly someone's
name would inadvertently be omitted. Those persons involved have the project's
heartfelt thanks for an impossible job well done.

The following projects and/or persons are thanked for their willingness to
share pupil materials, evaluation materials, and other ideas.

    Don Fineran, Mathematics Consultant, Oregon Department of Education
    Frank Lester, Indiana University
    Steve Meiring, Mathematics Consultant, Ohio Department of Education
    Harold Schoen, University of Iowa
    Iowa Problem Solving Project, Earl Ockenga, Manager
    Math Lab Curriculum for Junior High, Dan Dolan, Director
    Mathematical Problem Solving Project, John LeBlanc, Director

CONTENTS

INTRODUCTION

What is PSM?

PROBLEM SOLVING IN MATHEMATICS is a program of problem-solving lessons and teaching techniques for grades 4–8 and (9) algebra. Each grade-level book contains approximately 80 lessons and a teacher's commentary with teaching suggestions and answer key for each lesson. *Problem Solving in Mathematics* is not intended to be a complete mathematics program by itself. Neither is it supplementary in the sense of being extra credit or to be done on special days. Rather, it is designed to be integrated into the regular mathematics program. Many of the problem-solving activities fit into the usual topics of whole numbers, fractions, decimals, percents, or equation solving. Each book begins with lessons that teach several problem-solving skills. Drill and practice, grade-level topics, and challenge activities using these problem-solving skills complete the book.

PROBLEM SOLVING IN MATHEMATICS is designed for use with all pupils in grades 4–8 and (9) algebra. At-grade-level pupils will be able to do the activities as they are. More advanced pupils may solve the problems and then extend their learning by using new data or creating new problems of a similar nature. Low achievers, often identified as such only because they haven't reached certain computational levels, should be able to do the work in PSM with minor modifications. The teacher may wish to work with these pupils at a slower pace using more explanations and presenting the material in smaller doses.

[Additional problems appropriate for low achievers are contained in the *Alternative Problem Solving in Mathematics* book. Many of the activities in that book are similar to those in the regular books except that the math computation and length of time needed for completion are scaled down. The activities are generally appropriate for pupils in grades 4–6.]

Why Teach Problem Solving?

Problem solving is an ability people need throughout life. Pupils have many problems with varying degrees of complexity. Problems arise as they attempt to understand concepts, see relationships, acquire skills, and get along with their peers, parents, and teachers. Adults have problems, many of which are associated with making a living, coping with the energy crisis, living in a nation with peoples from different cultural backgrounds, and preserving the environment. Since problems are so central to living, educators need to be concerned about the growth their pupils make in tackling problems.

What Is a Problem?

MACHINE HOOK-UPS

Input Number

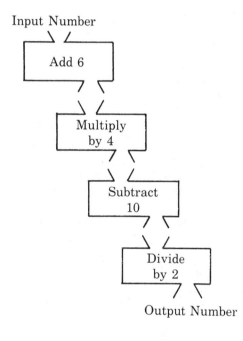

Output Number

It is highly recommended that teachers intending to use *Problem Solving in Mathematics* receive training in implementing the program. The *In-Service Guide* contains much of this valuable material. In addition, in-service audio cassette tapes are available. These provide indepth guidance on using the PSM grade-level books and an overall explanation of how to implement the whole program. The tapes are available for loan upon request. Please contact Dale Seymour Publications, Box 10888, Palo Alto, CA 94303 for further information about the tapes and other possible in-service opportunities.

	Input Number	Output Number
a.	4	
b.	8	
c.	12	
d.		39
e.		47
f.		61

Suppose a 6th grader were asked to fill in the missing output blanks for *a*, *b*, and *c* in the table. Would this be a problem for him? Probably not, since all he would need to do is to follow the directions. Suppose a 2nd-year algebra student were asked to fill in the missing input blank for *d*. Would this be a problem? Probably not, since she would write the suggested equation,

$$\frac{4\,(x\,+\,6)\,-\,10}{2} = 39$$

and then solve it for the input. Now suppose the 6th grader were asked to fill in the input for *d*, would this be a problem for him? Probably it *would* be. He has no directions for getting the answer. However, if he has the desire, it is within his power to find the answer. What might he do? Here are some possibilities:

1. He might make *guesses*, do *checking*, and then make refinements until he gets the answer.

2. He might fill in the output numbers that correspond to the input numbers for *a*, *b*, and *c*.

	Input Number	Output Number
a.	4	
b.	8	
c.	12	
d.		39

and then observe this pattern:
For an increase of 4 for the input, the output is increased by 8.
Such an observation should lead quickly to the required input of 16.

3. He might start with the output and *work backwards* through the machine hook-up using the inverse (or opposite) operations.

For this pupil, there was no "ready-made" way for him to find the answer, but most motivated 6th-grade pupils would find a way.

A *problem*, then, is a situation in which an individual or group accepts the challenge of performing a task for which there is no immediately obvious way to determine a solution. Frequently, the problem can be approached in many ways. Occasionally, the resulting investigations are nonproductive. Sometimes they are so productive as to lead to many different solutions or suggest more problems than they solve.

What Does Problem Solving Involve?

Problem solving requires the use of many *skills*. Usually these skills need to be used in certain combinations before a problem is solved. A combination of skills used in working toward the solution of a problem can be referred to as a *strategy*. A successful strategy requires the individual or group to generate the information needed for solving the problem. A considerable amount of creativity can be involved in generating this information.

What Problem-Solving Skills Are Used in PSM?

Skills are the building blocks used in solving a problem. The pupil materials in the PSM book afford many opportunities to emphasize problem-solving skills. A listing of these skills is given below.

THE PSM CLASSIFIED LIST OF PROBLEM-SOLVING SKILLS

A. Problem Discovery, Formulation
   1. State the problem in your own words.
   2. Clarify the problem through careful reading and by asking questions.
   3. Visualize an object from its drawing or description.
   4. Follow written and/or oral directions.

B. Seeking Information

  5. Collect data needed to solve the problem.

  6. Share data and results with other persons.

  7. Listen to persons who have relevant knowledge and experiences to share.

  8. Search printed matter for needed information.

  9. Make necessary measurements for obtaining a solution.

10. Record solution possibilities or attempts.

11. Recall and list related information and knowledge.

C. Analyzing Information

12. Eliminate extraneous information.

13. Find likenesses and differences and make comparisons.

14. Classify objects or concepts.

15. Make and use a drawing or model.

16. Make and/or use a systematic list or table.

17. Make and/or use a graph.

18. Look for patterns and/or properties.

19. Use mathematical symbols to describe situations.

20. Break a problem into manageable parts.

D. Solve—Putting It Together—Synthesis

21. Make predictions, conjectures, and/or generalizations based upon data.

22. Make decisions based upon data.

23. Make necessary computations needed for the solution.

24. Determine limits and/or eliminate possibilities.

25. Make reasonable estimates.

26. Guess, check, and refine.

27. Solve an easier but related problem. Study solution process for clues.

28. Change a problem into one you can solve. (Simplify the problem.)

29. Satisfy one condition at a time.

30. Look at problem situation from different points of view.

31. Reason from what you already know. (Deduce.)

32. Work backwards.

33. Check calculated answers by making approximations.

34. Detect and correct errors.

35. Make necessary measurements for checking a solution.

36. Identify problem situation in which a solution is not possible.

37. Revise the conditions of a problem so a solution is possible.

E. Looking Back—Consolidating Gains

38. Explain how you solved a problem.

39. Make explanations based upon data.

40. Solve a problem using a different method.

41. Find another answer when more than one is possible.

42. Double check solutions by using some formal reasoning method (mathematical proof).

43. Study the solution process.

44. Find or invent other problems which can be solved by certain solution procedures.

45. Generalize a problem solution so as to include other solutions.

F. Looking Ahead—Formulating New Problems

46. Create new problems by varying a given one.

What Are Some Examples of Problem-Solving Strategies?

Since strategies are a combination of skills, a listing (if it were possible) would be even more cumbersome than the list of skills. Examples of some strategies that might be used in the "Machine Hook-Ups" problem follow:

Strategy 1. *Guess* the input; *check* by computing the output number for your guess; if guess does not give the desired output, note the direction of error; *refine* the guess; compute; continue making refinements until the correct output results.

Strategy 2. *Observe* the *patterns* suggested by the input and output numbers for the a, b, c entries in the table; *predict* additional output and input numbers by extending both patterns; *check* the predicted input for the d entry by computing.

Strategy 3. *Study* the operations suggested in the machine hook-up; *work backwards* through the machine *using previous knowledge* about inverse operations.

An awareness of the strategies being used to solve a problem is probably the most important step in the development of a pupil's problem-solving abilities.

What is the Instructional Approach Used in PSM?

The content objectives of the lessons are similar to those of most textbooks. The difference is in the approach used. First, a wider variety of problem-solving skills is emphasized in the materials than in most texts. Second, different styles of teaching such as direct instruction, guided discovery, laboratory work, small-group discussions, nondirective instruction, and individual work all have a role to play in problem-solving instruction.

Most texts employ direct instruction almost exclusively, whereas similar lessons in PSM are patterned after a guided discovery approach. Also, an attempt is made in the materials to use intuitive approaches extensively before teaching formal algorithms. Each of the following is an integral part of the instructional approach to problem solving.

A. TEACH PROBLEM-SOLVING SKILLS DIRECTLY

Problem-solving skills such as "follow directions," "listen," and "correct errors" are skills teachers expect pupils to master. Yet, such skills as "guess and check," "make a systematic list," "look for a pattern," or "change a problem into one you can solve" are seldom made the object of direct instruction. These skills, as well as many more, need emphasis. Detailed examples for teaching these skills early in the school year are given in the commentaries to the *Getting Started* activities.

B. INCORPORATE A PROBLEM-SOLVING APPROACH WHEN TEACHING TOPICS IN THE COURSE OF STUDY

Drill and practice activities. Each PSM book includes many pages of drill and practice at the problem-solving level. These pages, along with the *Getting Started* section, are easy for pupils and teachers to get into and should be started early in the school year.

Laboratory activities and investigations involving mathematical applications and readiness activities. Readiness activities from such mathematical strands as geometry, number theory, and probability are included in each book. For example, area explorations are used in grade 4 as the initial stage in the teaching of the multiplication and division algorithms and fraction concepts.

Teaching mathematical concepts, generalizations, and processes. Each book includes two or more sections on grade-level content topics. For the most part, these topics are developmental in nature and usually need to be supplemented with practice pages selected from a textbook.

C. PROVIDE MANY OPPORTUNITIES FOR PUPILS TO USE THEIR OWN PROBLEM-SOLVING STRATEGIES

One section of each book includes a collection of challenge activities which provide opportunities for emphasizing problem-solving strategies. Generally, instruction should be nondirective, but at times suggestions may need to be given. If possible, these suggestions should be made in the form of alternatives to be explored rather than hints to be followed.

D. CREATE A CLASSROOM ATMOSPHERE IN WHICH OPENNESS AND CREATIVITY CAN OCCUR

Such a classroom climate should develop if the considerations mentioned in A, B, and C are followed. Some specific suggestions to keep in mind as the materials are used are:

- Set an example by solving problems and by sharing these experiences with the pupils.
- Reduce anxiety by encouraging communication and cooperation. On frequent occasions problems might be investigated using a cooperative mode of instruction along with brainstorming sessions.
- Encourage pupils in their efforts to solve a problem by indicating that their strategies are worth trying and by providing them with sufficient time to investigate the problem; stress the value of the procedures pupils use.
- Use pupils' ideas (including their mistakes) in solving problems and developing lessons.
- Ask probing questions which make use of words and phrases such as
  I wonder if . . . .
  Do you suppose that . . . .
  What happens if . . . .
  How could we find out . . . .
  Is it possible that . . . .
- Reinforce the asking of probing questions by pupils as they search for increased understanding. Pupils seldom are skilled at seeking probing questions but they can be taught to do so. If instruction is successful, questions of the type, "What should I do now?," will be addressed to themselves rather than to the teacher.

What Are the Parts of Each PSM Book?

PROBLEM SOLVING IN MATHEMATICS

Grade 4	Grade 5	Grade 6	Grade 7	Grade 8	Grade 9
Getting Started	Getting Started	Getting Started	Getting Started	Getting Started	Getting Started
Place Value Drill and Practice	Whole Number Drill and Practice	Drill and Practice	Drill and Practice- Whole Numbers	Drill and Practice	Algebraic Concepts and Patterns
Whole Number Drill and Practice	Story Problems	Story Problems	Drill and Practice- Fractions	Variation	Algebraic Explanations
Multiplication and Division Concepts	Fractions	Fractions	Drill and Practice- Decimals	Integer Sense	Equation Solving
Fraction Concepts	Geometry	Geometry	Percent Sense	Equation Solving	Word Problems
Two-digit Multiplication	Decimals	Decimals	Factors, Multiples, and Primes	Protractor Experiments	Binomials
Geometry	Probability	Probability	Measurement-Volume, Area, Perimeter	Investigations in Geometry	Graphs and Equations
Rectangles and Division	Estimation with Calculators	Challenges	Probability	Calculator	Graph Investigations
Challenges	Challenges		Challenges	Percent Estimation	Systems of Linear Equations
				Probability	Challenges
				Challenges	

Notice that the above chart is only a scope of PSM—not a scope and sequence. In general, no sequence of topics is suggested with the exceptions that *Getting Started* activities must come early in the school year and *Challenge* activities are usually deferred until later in the year.

Getting Started Several problem-solving skills are presented in the *Getting Started* section of each grade level. Hopefully, by concentrating on these skills during the first few weeks of school pupils will have confidence in applying them to problems that occur later on. In presenting these skills, a direct mode of instruction is recommended. Since the emphasis needs to be on the problem-solving skill used to find the solution, about ten to twelve minutes per day are needed to present a problem.

Drill And Practice No sequence is implied by the order of activities included in these sections. They can be used throughout the year but are especially appropriate near the beginning of the year when the initial chapters in the textbook emphasize review. Most of the activities are not intended to develop any particular concept. Rather, they are drill and practice lessons with a problem-solving flavor.

Challenges Fifteen or more challenge problems are included in each book. In general, these should be used only after *Getting Started* activities have been completed and pupils have had some successful problem-solving experiences.

Many of the other sections in PSM are intended to focus on particular grade-level content. The purpose is to provide intuitive background for certain topics. A more extensive textbook treatment usually will need to follow the intuitive development.

Teacher Commentaries Each section of a PSM book has an overview teacher commentary. The overview commentary usually includes some philosophy and some suggestions as to how the activities within the section should be used. Also, every pupil page in PSM has a teacher commentary on the back of the lesson. Included here are mathematics teaching objectives, problem-solving skills pupils might use, materials needed, comments and suggestions, and answers.

How Often Should Instruction Be Focused on Problem Solving?

Some class time should be given to problem solving nearly every day. On some days an entire class period might be spent on problem-solving activities; on others, only 8 to 10 minutes. Not all the activities need to be selected from PSM. Your textbook may contain ideas. Certainly you can create some of your own. Many companies now have published excellent materials which can be used as sources for problem-solving ideas. Frequently, short periods of time should be used for identifying and comparing problem-solving skills and strategies used in solving problems.

How Can I Use These Materials When I Can't Even Finish What's in the Regular Textbook?

This is a common concern. But PSM is not intended to be an "add-on" program. Instead, much of PSM can replace material in the textbook. Correlation charts can be made suggesting how PSM can be integrated into the course of study or with the adopted text. Also, certain textbook companies have correlated their tests with the PSM materials.

Can the Materials Be Duplicated?

The pupil lessons may be copied for students. Each pupil lesson may be used as an overhead projector transparency master or as a blackline duplicator master. Sometimes the teacher may want to project one problem at a time for pupils to focus their attentions on. Other times, the teacher might want to duplicate a lesson for individual or small group work. Permission to duplicate pupil lesson pages for classroom use is given by the publisher.

How Can a Teacher Tell Whether Pupils Are Developing and Extending Their Problem-Solving Abilities?

Presently, reliable paper and pencil tests for measuring problem-solving abilities are not available. Teachers, however, can detect problem-solving growth by observing such pupil behaviors as

- identifying the problem-solving skills being used.

- giving accounts of successful strategies used in working on problems.
- insisting on understanding the topics being studied.
- persisting while solving difficult problems.
- working with others to solve problems.
- bringing in problems for class members and teachers to solve.
- inventing new problems by changing problems previously solved.

What Evidence Is There of the Effectiveness of PSM?

Although no carefully controlled longitudinal study has been made, evaluation studies do indicate that pupils, teachers, and parents like the materials. Scores on standardized mathematics achievement tests show that pupils are registering greater gains than expected on all parts of the test, including computation. Significant gains were made on special problem-solving skills tests which were given at the beginning and end of a school year.

Also, when selected materials were used exclusively over a period of several weeks with 6th-grade classes, significant gains were made on the word-problem portion of the standardized test. In general, the greater gains occurred in those classrooms where the materials were used as specified in the teacher commentaries and in-service materials.

Teachers have indicated that problem-solving skills such as *look for a pattern, eliminate possibilities*, and *guess and check* do carry over to other subjects such as Social Studies, Language Arts, and Science. Also, the materials seem to be working with many pupils who have not been especially successful in mathematics. And finally, many teachers report that PSM has caused them to make changes in their teaching style.

Why Is It Best to Have Whole-Staff Commitment?

Improving pupils' abilities to solve problems is not a short-range goal. In general, efforts must be made over a long period of time if permanent changes are to result. Ideally, then, the teaching staff for at least three successive grade levels should commit themselves to using PSM with their pupils. Also, if others are involved, this will allow for opportunities to plan together and to share experiences.

How Much In-Service Is Needed?

A teacher who understands the meaning of problem solving and is comfortable with the different styles of teaching it requires could get by with self in-service by carefully studying the section and page commentaries in a grade-level book. The different styles of teaching required include direct instruction, guided discovery, laboratory work, small group instruction, individual work, and nondirective instructions. The teacher would find the audio tapes for each book and the *In-Service Guide* a valuable resource and even a time saver.

If a school staff decides to emphasize problem solving in all grade levels where PSM books are available, in-service sessions should be led by someone who has used the materials in the intended way. For more information on this in-service see the *In-Service Guide*.

What Materials Are Needed?

PROBLEM-SOLVING PROGRAM

REQUIRED MATERIALS	Grade 4	5	6	7	8	9
blank cards	X	X	X	X	X	X
bottle caps or markers	X			X		
calendar						X
calculators (optional for some activities)	X	X	X	X	X	X
cm squared paper, strips and singles						X
coins				X	X	
colored construction paper (circle fractions)	X	X				
cubes	X	X	X	X		X
cubes with red, yellow and green faces					X	
Cuisenaire rods (orange and white) or strips of paper		X				
dice (blank wooden or foam, for special dice)	X					
dice, regular (average 2 per student)	X	X	X	X	X	
geoboards, rubber bands, and record paper	X		X			
graph paper or cm squared paper				X		X
grid paper (1")			X			
metric rulers				X	X	X
phone books, newspapers, magazines		X		X		
protractors and compasses					X	
scissors	X	X	X	X		
spinners (2 teacher-made)			X			
tangrams	X					
tape measures				X		
thumbtacks (10 per pair of students)		X				
tile	X		X			
tongue depressors	X					
uncooked spaghetti or paper strips			X			

PSM   Rev. 1982

RECOMMENDED MATERIALS	Grade 4	5	6	7	8	9
adding machine tape				X		
centimetre rulers			X	X		
colored pens, pencils, or crayons		X				
coins, toy or real	X					
coins (two and one-half)						X
cubes					X	
demonstration ruler for overhead	X	X				
dominoes					X	
geoboard, transparent (for overhead)	X		X			
money - 20 $1.00 bills per student			X			
moveable markers	X	X	X		X	
octahedral die for extension activity				X		
overhead projector	X	X	X	X	X	X
place value frame and markers			X			
straws, uncooked spaghetti, or toothpicks		X				
transparent circle fractions for overhead	X	X				

Where Can I Find Other Problem Solving Materials?

RESOURCE BIBLIOGRAPHY

The number in parentheses refers to the list of publishers on the next page.

For students and teachers:

AFTER MATH, BOOKS I—IV by Dale Seymour, et al.
    Puzzles to solve -- some of them non-mathematical.   (1)

AHA, INSIGHT by Martin Gardner
    Puzzles to solve -- many of them non-mathematical.   (3)

THE BOOK OF THINK by Marilyn Burns
    Situations leading to a problem-solving investigation.   (1)

CALCULATOR ACTIVITIES FOR THE CLASSROOM, BOOKS 1 & 2 by George Immerzeel and
    Earl Ockenga
    Calculator activities using problem solving.   (1)

GEOMETRY AND VISUALIZATION by Mathematics Resource Project
    Resource materials for geometry.   (1)

GOOD TIMES MATH EVENT BOOK by Marilyn Burns
    Situations leading to a problem-solving investigation.   (1)

FAVORITE PROBLEMS by Dale Seymour
    Problem solving challenges for grades 5-7.   (3)

FUNTASTIC CALCULATOR MATH by Edward Beardslee
    Calculator activities using problem solving.   (4)

I HATE MATHEMATICS! BOOK by Marilyn Burns
    Situations leading to a problem solving investigation.   (3)

MATHEMATICS IN SCIENCE AND SOCIETY by Mathematics Resource Project
    Resource activities in the fields of astronomy, biology, environment,
    music, physics, and sports.   (1)

MIND BENDERS by Anita Harnadek
    Logic problems to develop deductive thinking skills. Books A-1, A-2, A-3,
    and A-4 are easy. Books B-1, B-2, B-3, and B-4 are of medium difficulty.
    Books C-1, C-2, and C-3 are difficult.   (6)

NUMBER NUTZ (Books A, B, C, D) by Arthur Wiebe
    Drill and practice activities at the problem solving level.   (2)

NUMBER SENSE AND ARITHMETIC SKILLS by Mathematics Resource Project
    Resource materials for place value, whole numbers, fractions, and decimals.   (1)

The Oregon Mathematics Teacher (magazine)
    Situations leading to a problem solving investigation.   (8)

PROBLEM OF THE WEEK by Lyle Fisher and William Medigovich
    Problem solving challenges for grades 7-12.   (3)

RATIO, PROPORTION AND SCALING by Mathematics Resource Project
    Resource materials for ratio, proportion, percent, and scale drawings.   (1)

STATISTICS AND INFORMATION ORGANIZATION by Mathematics Resource Project
    Resource materials for statistics and probability.   (1)

SUPER PROBLEMS by Lyle Fisher
        Problem solving challenges for grades 7-9.   (3)

For teachers only:

DIDACTICS AND MATHEMATICS by Mathematics Resource Project   (1)

HOW TO SOLVE IT by George Polya   (3)

MATH IN OREGON SCHOOLS by the Oregon Department of Education   (9)

PROBLEM SOLVING: A BASIC MATHEMATICS GOAL by the Ohio Department of Education   (3)

PROBLEM SOLVING: A HANDBOOK FOR TEACHERS by Stephen Krulik and Jesse Rudnik   (1)

PROBLEM SOLVING IN SCHOOL MATHEMATICS by NCTM   (7)

Publisher's List

1. Creative Publications, 3977 E Bayshore Rd, PO Box 10328, Palo Alto, CA 94303

2. Creative Teaching Associates, PO Box 7714, Fresno, CA 93727

3. Dale Seymour Publications, PO Box 10888, Palo Alto, CA 94303

4. Enrich, Inc., 760 Kifer Rd, Sunnyvale, CA 94086

5. W. H. Freeman and Co., 660 Market St, San Francisco, CA 94104

6. Midwest Publications, PO Box 448, Pacific Grove, CA 93950

7. National Council of Teachers of Mathematics, 1906 Association Dr, Reston, VA
    22091

8. Oregon Council of Teachers of Mathematics, Clackamas High School,
    13801 SE Webster St, Milwaukie, OR 97222

9. Oregon Department of Education, 700 Pringle Parkway SE, Salem, OR 97310

Algebra

I. GETTING STARTED

I. GETTING STARTED

Teachers usually are successful at teaching skills in mathematics. Besides computation and algebra skills, they emphasize skills in following directions, listening, detecting errors, explaining, recording, comparing, measuring, sharing, ... They (You!) can also teach problem-solving skills. This section is designed to help teachers teach and pupils learn specific problem-solving skills.

I'm good at teaching skills -- bet I could teach my pupils these problem-solving skills!

Some Problem-Solving Skills

Five common but powerful problem-solving skills are introduced in this section. They are:
. guess and check
. look for a pattern
. make a systematic list
. make and use a drawing or model
. simplify the problem.

Pupils <u>might</u> use other skills to solve the problems. They can be praised for their insight, but it is usually a good idea to limit the emphasized list of skills directly taught during the first few lessons. More problem-solving skills will occur in the other sections.

An Important DON'T

When you read the episodes that follow in this <u>Getting Started</u> section, notice how the lessons are <u>very teacher directed</u>. The main purpose is to teach the problem-solving skills. Teachers should stress the skills verbally and write them on the board. <u>Don't</u> just ditto these activities and hand them out to be worked. Teacher direction through questions, summaries, praise, etc., is <u>most</u> important for teaching the problem-solving skills in this section. We want pupils to focus on specific skills which will be used often in all the sections. Later, in the <u>Challenges</u> section, pupils will be working more independently.

Using The Activities

If you heed the important <u>Don't</u> on the previous page, you are on your way to success! The problems here should fit right in with your required course of study as they use whole number and fraction skills, elementary geometry, and money concepts. In most cases, pupils will have the prerequisites for the problems in this section although you might want to check over each problem to be sure.

No special materials are required. The large type used for the problems makes them easier to read if they are shown on an overhead screen. In most cases pupils can easily copy the problem from the overhead. At other times, you might copy the problem onto the chalkboard.

When And How Many

The <u>Getting Started</u> section should be used at the beginning of the year as it builds background in problem-solving skills for the other sections. As the format indicates, <u>only one problem per day</u> should be used. Each should take less than twelve minutes of classtime if the direct mode of instruction is used. The remainder of the period is used for a lesson from the textbook or perhaps an activity from the <u>Patterns And Concepts</u> section of these materials.

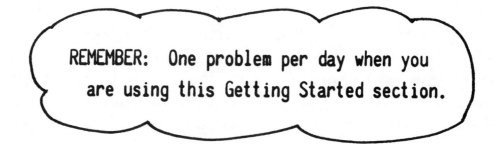

REMEMBER: One problem per day when you are using this Getting Started section.

Guess And Check

The episode that follows shows how one teacher teaches the skill of guess and check. Notice how she very closely directs the instruction and constantly uses the terminology.

It is near the beginning of the year and Ms. Vaughn is about to start a math lesson. After getting the attention of the class, she begins.

Ms. V: (Reads from overhead.) Matthew is 12 years old and his mother is 3 times as old. How many years must pass before his mother is twice as old? We can find the answer by guessing. Is the answer 38 years?

Mark: No, that's not correct.

Ms. V: Why not?

Mark: His mother is 36 now. In 38 years Matthew will be 50 and she will be 74. To be twice as old she would have to be 100 years old.

Ms. V: Good. You checked my guess and found it was too high. Well, how about 4 years from now as a guess? (Pause)

Tammy: That won't work either. Matthew will be 16 and the mother will be 40 which is $2\frac{1}{2}$ times as much.

Ms. V: Correct, you found that guess to be too small. The correct guess must be between 4 and 50. How about a guess at 10 years from now?

Mike: That's close--22 years and 46 years is almost twice as much. It must be 12 years.

Ms. V: Yes, Mike. We used several guesses to narrow in on the correct answer. Now let's try the second part of the problem.

Ms. V: (Reads from overhead.) Matthew is 12 years old and his mother is 3 times as old. How many years ago was she 7 times as old. I wonder if it was 3 years ago. Is it?

Ben: No.

Ms. V: How do you know?

Ben: His mother is 36 and 3 years ago they were 9 and 33.

Ms. V: By checking my guess, you found it was off. Could it be 10 years ago? June?

June: They were 2 and 26 ten years ago. That's 13 times older.

Ms. V: Ten years ago she was 13 times older; 3 years ago she was between 3 and 4 times older. Can you refine your guesses to get closer?

Barb: Try 6. Let's see, that's 6 and 30--5 times--not enough.

Joe: Maybe 8. That gives 4 and 28. That's right.

Ms. V: Good. You found the correct answer of 8 years ago. Guess and check is a good way to solve problems. We're going to use it throughout the year. I'm going to write it up on a poster so we'll all remember how important it is!

GUESS AND CHECK

WEEK 1 - DAY 1

Matthew is 12 years old and his mother is 3 times as old.
How many years must pass before his mother is twice as old?
How many years ago was she 7 times as old?

****

WEEK 1 - DAY 2

Toni has exactly $2.00 in nickels and dimes. She has twice
as many dimes as nickels. How many of each does she have?

****

WEEK 1 - DAY 3

Penny bought a scarf for $5.00, spent $\frac{1}{2}$ of her remaining
money on jogging shoes, bought lunch for $2.00, then spent $\frac{1}{2}$
of her remaining money on a present. She had $10 left.
How much did she start with?

Guess And Check

Day 1. Answer:

12 years from now they will be 24 and 48, twice as old.
8 years ago they were 4 and 28, seven times as old.

Comments and suggestions:

· See the introductory commentary.

Day 2. Answer:

16 dimes, 8 nickels

Comments and suggestions:

. Start with a guess of 10 dimes and 5 nickels which gives $1.25.
  This is not enough. A guess of 20 dimes and 10 nickels gives $2.50
  which is too much. Pupils can quickly see that the number of dimes
  must be an even number between 10 and 20.

Day 3. Answer:

$49

Comments and suggestions:

. Start with a guess of $85. Working through the problem gives an
  ending amount of $19 which is too much. The guess is too high.
. Guess $25 which gives $4 which is not enough. The guess is too low.
. Guess $45 which gives $9. The guess is too low but very close to the
  correct answer of $49.
. Some pupils might want to work backwards from the $10. Although
  appropriate at a later time, emphasize the guess and check skill at
  this time.

Guess And Check (cont.)

WEEK 1 - DAY 4

    a.   I'm thinking of a number. If you multiply it by 3, then subtract 5 and finally add 10, you get 20. What number am I thinking of?

    b.   I'm thinking of a number. If you multiply its square by 3 and then add 9, you get 117. What is that number?

    c.   I'm thinking of a number. If you subtract 4 from the number, then multiply the result by 3 and then add 5, you get 26. What is that number?

****

WEEK 1 - DAY 5

. Each check I write costs 10¢. I also have to pay a flat fee of 25¢ per month.

. My bank sent me a letter saying the fees were going to change.

. The new fee is 8¢ for each check and a flat fee of 50¢ a month.

. The bank said this would be cheaper for me because of the number of checks I write.

. What is the least number of checks I must write to make the new fees cheaper?

Day 4. Answers:

       a. 5   b. 6   c. 11

Comments and suggestions:

. Each of these might be solved by the work backwards skill. Although
appropriate at a later time, emphasize guess and check now.

. Suggested guesses for each of the problems are
    a. 10 (too high), 2 (too low)
    b. 10 (too high), 5 (too low)
    c. 10 (too low), 15 (too high)

Day 5. Answer:

13 checks

Comments and suggestions:

. A guess of 10 checks gives an original cost of $1.25 and a new cost
of $1.30, not enough checks to make the new cost less.

. For a guess of 15, we have an old cost of $1.75 and a new cost of
$1.70. That's less, but is it the least number?

. Twelve checks cost $1.45 and $1.46, not enough. Thirteen checks gives
$1.55 and $1.54. Thirteen is the least number of checks.

<u>Look For A Pattern</u>

Much of algebra is looking for and applying useful patterns. Here are two cases where patterns are commonly used to motivate meaning in algebra.

a. Identify the pattern in these products, then predict products for -1 X 3 and -2 X 3.

$$4 \times 3 = 12$$
$$3 \times 3 = 9$$
$$2 \times 3 = 6 \qquad \text{(Subtract 3 from}$$
$$1 \times 3 = 3 \qquad \text{each preceding}$$
$$0 \times 3 = 0 \qquad \text{product.)}$$
$$-1 \times 3 =$$
$$-2 \times 3 =$$

b. Identify the pattern, then use the pattern to predict values for 2^0 and 2^{-1}.

$$2^4 = 16$$
$$2^3 = 8 \qquad \text{(Take one-half}$$
$$2^2 = 4 \qquad \text{of each preceding}$$
$$2^1 = 2 \qquad \text{product.)}$$
$$2^0 =$$
$$2^{-1} =$$

Because patterns are so important in algebra, Mr. Stone introduced this problem-solving skill very early. Notice that the problem he used also reviews some basic ideas such as odd numbers and squares.

Mr. S: Who remembers what an odd number is?

Jeff: It's like 1, 3, 5, ... not divisible by 2.

Mr. S: That's right. I want you to find these sums. I'll write up the answers as you get them.

What is the sum of the first 4 odd numbers?

first 5 odd numbers?   (on board)

first 7 odd numbers?

first 10 odd numbers?

Polly: The first one is 16.

Ned: The sum of the first five is 25.

Millie: The next one is 49. Do we have to work the next one?

Mr. S: Sure--I want to see if you remember how to add after summer vacation!

Frank: It's 100.

Mr. S: Good. Now I've got a tough one for you. Find the sum of the first 3467 odd numbers.

Jim: You've got to be kidding.

Mr. S: No. I'm very serious. However, this time you don't have to add them all up. There's an easier way if you see a pattern in the results on the board. Keep your answers quiet to give everyone a chance.

Sara: It seems impossible.

Sally: I think I've got it!

Joe: Let me see if you got what I did.

Mr. S: (After others think they have it.) Okay, what pattern did you see?

Joe: They are all squares. However many you add up, just find the
           square of that number.

Mr. S: Good work, class. Looking for a pattern is a very important
           problem solving skill. We'll use it all year.

Look For A Pattern (cont.)

LOOK FOR A PATTERN

What is the sum of the first 4 odd numbers?

5 odd numbers?

7 odd numbers?

10 odd numbers?

3467 odd numbers?

****

Find the pattern. Fill in the blanks.

a. 1, 4, 9, 16, 25, ___, ___, ___

b. 3, 4, 7, 11, 18, 29, ___, ___, ___

c. 5, 10, 9, 18, 17, 34, 33, ___, ___

d. 1, 2, 6, 24, 120, ___, ___, ___

e. 77, 49, 36, 18, ___

****

Find these products:

7 x 9 = _____

77 x 99 = _____

777 x 999 = _____

Predict the problem for this product: ____ x ____ = 77,762,223
Check with a calculator.
Predict the product for this problem: 77,777 x 99,999 = _____

Look For A Pattern

Day 1. Answers:

16, 25, 49, 100; The square of 3467 or 3467^2 or 12,020,089

Comments and suggestions:

. See the introductory commentary.

Day 2. Answers:

    a. 36; 49; 64 – successive squares.

    b. 47; 76; 123 – add previous two terms.

    c. 66; 65 – double, then subtract 1.

    d. 720; 5040; 40,320 – multiply by successive counting numbers.

    e. 8 – multiply the digits of the number together.

Comments and suggestions:

. Be accepting of other patterns found by pupils. For example, one might
  see c as a pattern of odd, even, odd, even. An answer to e might be
  15, if the pupil sees successive differences as 28, 13, 18, 3.

. Problem e is unlike most patterns pupils have seen. A large hint of
  "Think multiplication" may be necessary.

Day 3. Answers

63; 7623; 776,223; 7,777 x 9,999; 7,777,622,223

Comments and suggestions:

. A calculator is helpful until it overflows.

. Some pupils will need additional help seeing that the number of 7s and
  2s in the final product is one less than the number of 7s and 9s in
  the problem.

Look For A Pattern (cont.)

WEEK 2 – DAY 4

For each problem, give the missing numbers or expressions.

a.

1	2	3	4	5	8	10	
3	6	9	12				48

b.

1	2	3	4	5	9	12	
3	5	7	9				37

c.

1	2	3	4	5	8	
$\frac{1}{2}$	$\frac{1}{6}$	$\frac{1}{12}$	$\frac{1}{20}$			$\frac{1}{110}$

d.

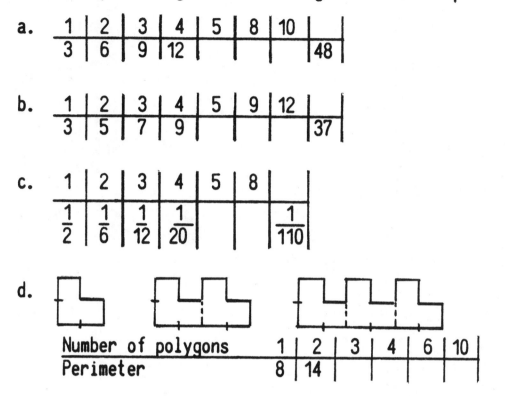

Number of polygons	1	2	3	4	6	10
Perimeter	8	14				

****

WEEK 2 – DAY 5

Fill out the table below.

Power of 2	2^1	2^2	2^3	2^4	2^5	2^6	2^7	2^8
Units digit	2	4						6

Predict the units digit for these:

2^{12}	2^{36}	2^9	2^{25}	2^{39}	2^{42}

Day 4.   Answers:

a.

5	8	10	16
15	24	30	48

(Multiply by 3.)

b.

5	9	12	18
11	19	25	37

(Double and add 1.)

c.

5	8	10
$\frac{1}{30}$	$\frac{1}{72}$	$\frac{1}{110}$

(Multiply number by next higher number; put answer under 1.)

d

3	4	6	10
20	26	38	62

(Multiply by 6 and add 2.)

Comments and suggestions:

. Be accepting of other patterns found by pupils. For example, one
  might see b as "add 2 each time." An answer to c might be "multiply
  the two numbers at the top." An answer to d might be "add 6 each time."

Day 5.   Answers:

Power of 2	2^3	2^4	2^5	2^6	2^7	2^8
Units digit	8	6	2	4	8	6

2^{12}	2^{36}	2^9	2^{25}	2^{39}	2^{42}
6	6	2	2	8	4

Comments and suggestions:

. Some pupils will need review on the meaning of exponents.
. Pupils might need to extend the table to see the pattern.
. The pattern to see is
    if the exponent is a multiple of 4, the units digit is 6.
    If the exponent is even and is not a multiple of 4, the units
    digit is 4.   If the exponent is one more than a multiple of 4,
    the units digit is 2.   If the exponent is one less than a
    multiple of 4, the units digit is 8.
. An easier problem involving last digits is 9^n. (Or 4^n, 5^n, 6^n)

Often a systematic list or table can make the search for a solution much easier. In the example below, the organized table makes order out of chaos and pupils are able to see when they have exhausted the possibilities.

Karen made up a puzzle.

"Take 25 marbles. Put them in 3 piles so an odd number of marbles is in each pile."

How many different solutions does Karen's puzzle have?

Ms. James:	Here is a problem for you to work on. Does anyone want to guess the number of solutions for the puzzle?	
Gary:	I'd guess 6.	
Gina:	More than 10.	
Ms. J:	How can we find out how many there are?	
Lennie:	Write them down and count them.	
Ms. J:	Do you have a solution, Lennie?	
Lennie:	5, 5 and 15.	
Ms. J:	Good. I'll write the solutions on the board as you give them.	
Kim:	3, 5 and 17.	
Others:	7, 5 and 13; 1, 9 and 15; 9, 5 and 11.	
Cassie:	5, 7 and 13.	
Dennis:	You already have that one. See - there's 7, 5 and 13.	

Ms. J: We're losing track of which ones we have. Maybe we need to make a systematic list. Let's start with the smallest odd number in one pile and work up. Let's see, with 1 marble we can have 1, 1, 23 or 1, 3, 21 or 1, 5, 19. (Writes on board.)

John: 1, 7, 17 then 1, 9, 15 then 1, 11, 13 then 1, 13, 11.

Ginnie: Wait! That's the same one reversed. You're done with the 1's. Now do 3's.

(After the list is completed.)

Ms. J: Are you sure we have them all?

Manley: Yes. Look, if you try 9's next, then we start repeating ...

(Class discussion convinces most pupils the list is complete.)

Smallest Pile		Largest Pile
1	1	23
1	3	21
1	5	19
1	7	17
1	9	15
1	11	13
3	3	19
3	5	17
3	7	15
3	9	13
3	11	11
5	5	15
5	7	13
5	9	11
7	7	11
7	9	9

Ms. J: That makes 16 solutions. Do you see how a systematic list helped us to solve the problem? For the next few days we will emphasize the skill, make a systematic list.

MAKE A SYSTEMATIC LIST

WEEK 3 - DAY 1

Karen made up a puzzle:

Take 25 marbles. Put them in 3 piles so an odd number of marbles is in each pile.

How many different solutions does Karen's puzzle have?

****

WEEK 3 - DAY 2

A rectangle has an area of 120 cm^2. Its length and width are whole numbers. What are possibilities for the two numbers? Which possibility gives the smallest perimeter?

****

WEEK 3 - DAY 3

The product of two whole numbers is 96 and their sum is less than 30. What are possibilities for the two numbers?

****

WEEK 3 - DAY 4

Jamie and Lynn each worked a different number of days. But each earned the same amount of money. Use the clues below to find how many days each worked.

. Jamie earned $15 a day.

. Lynn earned $10 a day.

. Lynn worked 5 more days than Jamie.

Make A Systematic List

Day 1. Answers:

16 solutions

Smallest Pile		Largest Pile
1	1	23
1	3	21
1	5	19
1	7	17
1	9	15
1	11	13
3	3	19
3	5	17
3	7	15
3	9	13
3	11	11
5	5	15
5	7	13
5	9	11
7	7	11
7	9	9

Comments and suggestions:

. See the introductory commentary.

Day 2. Answers:

Width	1	2	3	4	5	6	8	10
Length	120	60	40	30	24	20	15	12

The 10 by 12 rectangle has the least perimeter.

Comments and suggestions:

. The problem serves as a good review for divisibility rules.

. Some pupils might argue for eight more answers where a 10 by 12 and a 12 by 10 are different rectangles. Let the class decide if they want this distinction made.

Day 3. Answers:

4 and 24; 6 and 16; 8 and 12.

Comments and suggestions:

. Pupils can list all the pairs of whole numbers whose product is 96, then check the sums.

Day 4. See page 22.

Make A Systematic List (cont.)

WEEK 3 – DAY 5

Lonnie has a large supply of quarters, dimes, nickels, and pennies. Show how she could make change for 50¢ in 49 different ways.

25¢	10¢	5¢	1¢

25¢	10¢	5¢	1¢

Day 4. Answer:

Lynn worked 15 days and Jamie worked 10 days.

Comments and suggestions:

. A list makes the solution clear.
  Since Lynn worked 5 more days, the list
  shows she would have made $50 before
  Jamie started working.

. It is interesting to note that both
  earned the same amount at $30, $60, $90,
  and $120, but at each stage Lynn had not
  worked 5 days longer.

Day	Lynn	Jamie
1	$ 10	
2	$ 20	
3	$ 30	
4	$ 40	
5	$ 50	
6	$ 60	$ 15
7	$ 70	$ 30
8	$ 80	$ 45
9	$ 90	$ 60
10	$100	$ 75
11	$110	$ 90
12	$120	$105
13	$130	$120
14	$140	$135
15	$150	$150

Day 5. Answers:

25¢	10¢	5¢	1¢
2	-	-	-
1	2	1	-
1	2	-	5
1	1	3	-
1	1	2	5
1	1	1	10
1	1	-	15
1	-	5	-
1	-	4	5
1	-	3	10
1	-	2	15
1	-	1	20
1	-	-	25
-	5	-	-
-	4	2	-
-	4	1	5
-	4	-	10
-	3	4	-
-	3	3	5
-	3	2	10
-	3	1	15
-	3	-	20
-	2	6	-
-	2	5	5
-	2	4	10

25¢	10¢	5¢	1¢
-	2	3	15
-	2	2	20
-	2	1	25
-	2	-	30
-	1	8	-
-	1	7	5
-	1	6	10
-	1	5	15
-	1	4	20
-	1	3	25
-	1	2	30
-	1	1	35
-	1	-	40
-	-	10	-
-	-	9	5
-	-	8	10
-	-	7	15
-	-	6	20
-	-	5	25
-	-	4	30
-	-	3	35
-	-	2	40
-	-	1	45
-	-	-	50

Comments and suggestions:

. The problem is an ex-
  cellent one for showing
  how quickly many answers
  can be generated if one
  has a systematic list
  and can develop a
  pattern for completing
  the list.

. To impress your pupils
  with the quickness that
  the answers can be
  written, you might want
  to practice a few times,
  first!

Make And Use A Drawing Or Model

Many problems become easier to solve when we make and use a drawing or model. Models and drawings are especially needed in algebra where the notation or word problems often become too abstract for students to follow.

Here is how one teacher introduced this skill:

Ms. Manley: Let's review how to find the volume of a rectangular solid-- that's a shoe-box-like shape. (Shows pictures on overhead and pupils find the volume.) Now that you all remember how to find the volume of a rectangular solid, here's a problem for you.

A 12 by 16 rectangular piece of paper has a 2 by 2 square cut out of each corner. Then the sides are folded up to make an open box. Find the volume of the box.

Mindy: (Who started computing immediately.) I've almost got it -- it's 384.

Wes: No it's not. It's 280.

Ms. M: Mindy, how did you get 384?

Mindy: I multiplied 16 X 12 then that by 2. (Ms. Manley writes 16 X 12 X 2 = 384)

Ms. M: And Wes? How did you get 280?

Wes: 10 times 14 times 2. I subtracted two from both numbers.

Ms. M: Mmm--Looks like we aren't sure how to use the given numbers. Let's see if a drawing will help make it clear. We started with a 12 by 16 rectangle and cut out 2 by 2 square corners. Does this look right? Does it help?

Tim: Now you can see the bottom of the box and it's 8 by 12.

Ms. M: (Draws in bottom of the box.) Good. Now what will its volume be when the sides are turned up?

Daphne: Wait, Ms. Manley. I still don't really understand the problem.

Ms. M: Oh. Maybe this will help. (She holds up a piece of paper, cuts out square corners and folds up the sides to show an open box.) When the corners are cut out we can fold up the sides. Notice that the bottom of the box is the rectangle left in the middle of the paper. Now, what is the volume of the box?

Daphne: 8 times 12 times 2. Let's see, that's 192.

Ms. M: Right. Do you see how the drawing clears up the problem? Make and use a drawing or model is a useful problem solving skill. We'll be using it all week.

MAKE AND USE A DRAWING OR MODEL

WEEK 4 – DAY 1

A 12 by 16 rectangular piece of paper has a 2 by 2 square cut out of each corner. Then the sides are folded up to make an open box. Find the volume of the box.

****

WEEK 4 – DAY 2

Four squares are arranged so each square touches at least one other square. Any two squares touch each other according to these rules:

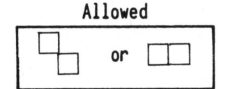
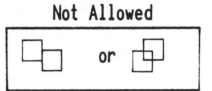

What are the possible perimeters?

****

WEEK 4 – DAY 3

Billy, Ray, Gary, Floyd, Clayton, and Pete just completed a 3000 metre race. Use these clues to help you determine the order in which they finished.

- . Floyd finished 10 seconds behind Gary.
- . Ray beat Pete by 20 seconds.
- . Billy finished 4 seconds behind Gary and 30 seconds ahead of Ray.
- . Clayton's finish was halfway between Floyd and Pete.

Day 1. Answer:

192

Comments and suggestions:
See the introductory commentary.

Day 2. Answers:

8   10   12   14   16

Comments and suggestions:

. Provide four tiles (squares) for pupils to manipulate.

. Some possible drawings corresponding to the perimeters above are:

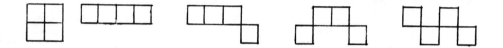

Day 3. Answer:

Gary first, Billy second, Floyd third, Clayton fourth, Ray fifth, and
Pete sixth.

Comments and suggestions:

. It is helpful to place Floyd and Gary to the far right on a number
  line with 10 units between them. The third clue can then be used
  putting Billy 4 seconds behind Gary and marking Ray 30 units behind
  Billy. The time between Pete and Floyd must then be calculated
  (it is 44 seconds) and Clayton can be placed on the line just ahead
  of Ray.

Make And Use A Drawing Or Model (cont.)

WEEK 4 - DAY 4

A 9 metre by 12 metre rectangular garden has a walk one metre wide all around it. What is the area of the walk?

****

WEEK 4 - DAY 5

One painter can paint a wall in ten minutes. Another painter can paint it in six minutes. About how long will it take both painters to do the job? Use the drawing below to help.

Day 4.   Answer:

46 m^2

Comments and suggestions:

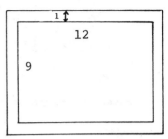

. Some pupils may believe the answer is
  obtained by multiplying 10 and 13 or
  11 and 14. Others may think of adding
  9 + 9 + 12 + 12.

. A drawing should help them see the
  different rectangles whose areas must
  be added.

. Correct answers can be obtained by several methods.

9 + 9 + 12 + 12 + 4 (1)

or by using 10 + 13 + 10 + 13 from

or (14 x 11) - (12 x 9)

Day 5.   Answer:

About 4 minutes

Comments and suggestions:

. You recognize this as a word problem. The drawing shows a wall marked
  in sixths on the outside and tenths on the inside to correspond to the
  minutes needed to paint the wall.

. The sequence of drawings below shows how an answer is obtained.

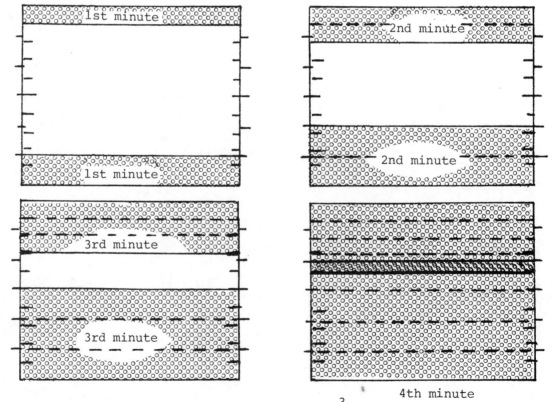

. For you only--the algebra works out to be $3\frac{3}{4}$ minutes.

Simplify the Problem

Sometimes applying a suitable problem solving skill can give insight into a seemingly unsolvable problem. This teacher used a difficult problem to impress pupils with the power of the skill simplify the problem. Warning: You might want to start with one of the other problems depending on your class.

Mr. Sloan: Suppose we each have a bag of pennies. We take turns putting them on a rectangular table. No penny can hang over the edge and no pennies can overlap although they can touch. Whoever puts a penny in the last space available wins all the pennies. Would you want to play first or second? How would you play to win? (Waits--pupils look puzzled.)

Jeff: It doesn't seem possible to figure out.

Mr. S: It does seem overwhelming. What if we had a one penny table? (Shows square and pennies on overhead.)

Pat: I'd play first and win.

Mr. S: Well, that was easy. What about a two-penny sized table?

Mollie: I'd play second.

Mr. S: Okay, I'll play first and I choose to put my penny in the middle.

Mollie: Oh--I should have played first in the middle.

Mr. S: Here's a 3-penny table.

Cam: I'll play first in the middle.

Mr. S: Then I'm stuck with playing on one side and you play on the other to win. Now, how about a four-penny table?

Tina: I'd go with the first penny in the middle. Then I'm guaranteed to win.

      or

Simplify the Problem

Mr. S: Now what if we had a bigger table, say a six-penny sized table.

Sam: Play first in the middle again.

Mr. S: I'll play in the corner.

Sam: I'll play below you. No! Wait! You could play in the middle of the other side. I'll play opposite you on the diagonal.

Mr. S: You are all catching on. Let's see. It seems it is best to play in the exact middle of the table. Then play opposite my play across the middle. Do you see how it helped to simplify the problem? That is our new problem solving skill for the week.

SIMPLIFY THE PROBLEM

WEEK 5 - DAY 1

Suppose we each have a bag of pennies. We take turns putting them on a rectangular table. No penny can hang over the edge and no pennies can overlap although they can touch. Whoever puts a penny in the last space available wins all the pennies. Would you want to play first or second? How would you play to win?

****

WEEK 5 - DAY 2

How many line segments can be drawn between each of 15 points? No three of the points are in a straight line.

****

WEEK 5 - DAY 3

Find the product:

$$(1 - \frac{1}{2})(1 - \frac{1}{3})(1 - \frac{1}{4}) \ldots (1 - \frac{1}{98})(1 - \frac{1}{99})(1 - \frac{1}{100}) =$$

<u>Simplify</u> <u>The</u> <u>Problem</u>

Day 1. Answers:

To assure a win, play first in the middle. After that, match your opponent's play by playing symmetrically across the middle. You will always be guaranteed a space until you put in the last penny.

Comments and suggestions:

. See the introductory commentary.

Day 2. Answer:

105 line segments

Comments and suggestions:

. To simplify the problem, consider the number of line segments with just 2, 3, 4, etc. points. The information forms an interesting list which leads to the answer.

Points	2	3	4	5	6	7	8	9	10	11	12	13	14	15
Line segments	1	3	6	10	15	21	28	36	45	55	66	78	91	105

. This problem has other solution strategies. At this point, emphasize the "simplify the problem" skill.

Day 3. Answer:

$$\frac{1}{100}$$

Comments and suggestions:

. Simplify the problem by considering the following cases:

$$(1 - \frac{1}{2}) = \frac{1}{2} = \frac{1}{2}$$

$$(1 - \frac{1}{2})(1 - \frac{1}{3}) = \frac{1}{2} \times \frac{2}{3} = \frac{1}{3}$$

$$(1 - \frac{1}{2})(1 - \frac{1}{3})(1 - \frac{1}{4}) = \frac{1}{2} \times \frac{2}{3} \times \frac{3}{4} = \frac{1}{4}$$

. The final product will be 1 over the last denominator.

Simplify The Problem (cont.)

Find the sum:

$$\frac{1}{1 \cdot 2} + \frac{1}{2 \cdot 3} + \frac{1}{3 \cdot 4} + \ldots + \frac{1}{98 \cdot 99} + \frac{1}{99 \cdot 100} =$$

****

Ten strangers attend a meeting. As introductions are made, each person shakes hands with all the others. How many handshakes occur?

Day 4. Answer:

$$\frac{99}{100}$$

Comments and suggestions:

. Simplify the problem by considering the following cases:

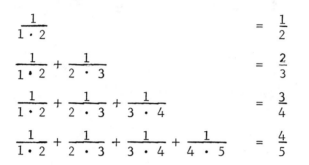

$$\frac{1}{1 \cdot 2} = \frac{1}{2}$$

$$\frac{1}{1 \cdot 2} + \frac{1}{2 \cdot 3} = \frac{2}{3}$$

$$\frac{1}{1 \cdot 2} + \frac{1}{2 \cdot 3} + \frac{1}{3 \cdot 4} = \frac{3}{4}$$

$$\frac{1}{1 \cdot 2} + \frac{1}{2 \cdot 3} + \frac{1}{3 \cdot 4} + \frac{1}{4 \cdot 5} = \frac{4}{5}$$

. The final sum will be found in the denominator of the last fraction.

Day 5. Answer:

45 handshakes

Comments and suggestions:

. To simplify the problem, consider the number of handshakes with just
  2, 3, 4, etc. persons. The information forms an interesting list
  which leads to the answer.

Persons	2	3	4	5	6	7	8	9	10
Handshakes	1	3	6	10	15	21	28	36	45

. This problem has other solution strategies. At this point, emphasize
  the "simplify the problem" skill.

. Point out how this problem is analogous to the problem of drawing
  line segments between non-collinear points.

Algebra

II. ALGEBRAIC CONCEPTS AND PATTERNS

II. ALGEBRAIC CONCEPTS AND PATTERNS

A variety of lessons are included in this section--introduction to units,
egrating new ideas with old ones, motivational materials including number
iosities, and practice activities. Collectively, all of them provide
ortunities for developing problem-solving skills such as generalizing a
olem to include other solutions, creating new problems by varying a given
, making explanations based upon data, using mathematics symbols to describe
ations, and working backwards. Activities are to be used to supplement the
elopment of units from textbooks or prescribed by courses of study.

These activities do not have to be done in the order in which they appear.
it is important they be used at the most appropriate time. The listing
ch follows should assist you in this task.

son Titles and a Brief Description

STRIPS AND SINGLES

1. Match the arithmetic expression with the geometric figure
    it describes. Each strip is ten units long.

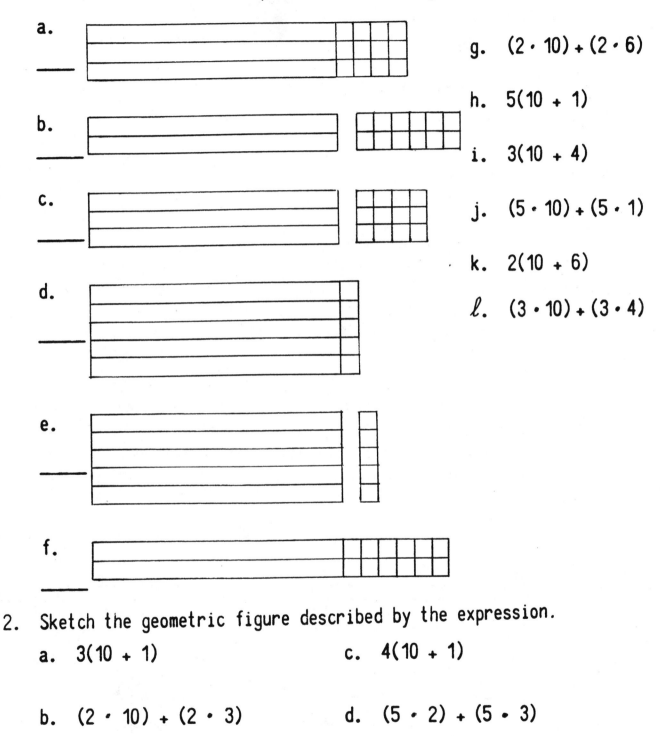

a. ____

b. ____

c. ____

d. ____

e. ____

f. ____

g. $(2 \cdot 10) + (2 \cdot 6)$

h. $5(10 + 1)$

i. $3(10 + 4)$

j. $(5 \cdot 10) + (5 \cdot 1)$

k. $2(10 + 6)$

ℓ. $(3 \cdot 10) + (3 \cdot 4)$

2. Sketch the geometric figure described by the expression.

a. $3(10 + 1)$

b. $(2 \cdot 10) + (2 \cdot 3)$

c. $4(10 + 1)$

d. $(5 \cdot 2) + (5 \cdot 3)$

Strips And Singles

Mathematics teaching objectives:

. Visualize and illustrate the distributive property using geometric figures

. Use the distributive property in calculations.

Problem-solving skills pupils <u>might</u> use:

. Make and use a drawing.

. Look for patterns.

. Solve an easier but related problem.

Materials needed:

. None

Comments and suggestions:

. The distributive property is introduced using drawings and pupils' arith-metic knowledge.

. Use this lesson along with the next one, <u>Algebraic</u> <u>Strips</u> <u>And</u> <u>Singles</u>, early in the year when introducing the use of variables. Both lessons can be done during the same class period. The second lesson is an extension of the first.

. Assist in the interpretation of the drawings by working exercises 1 and 2 in class.

. Compare the equivalent expressions in the examples of the distributive property. They use the same numbers and both involve the operations of multiplication and addition. Suggest that pupils explore the possibility of a similar property involving a different combination of operations.

. Point out ways the distributive property can be used in mental computation

Answers:

1. a. <u>i</u>　　b. <u>g</u>　　c. <u>l</u>　　d. <u>h</u>　　e. <u>j</u>　　f. <u>k</u>

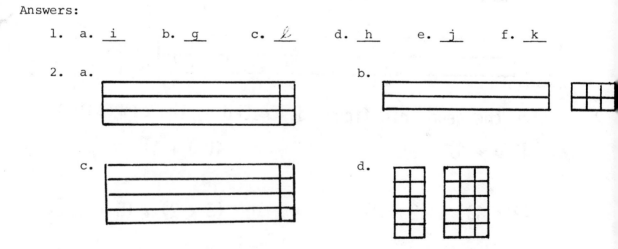

3. Match problems which have the same answers. Not all of them can be matched.

a. 3(5 + 17)

A. 5(4 + 8)

b. 10(8 + 4)

B. (150 · 29) + (150 · 62)

c. (5 · 4) + (5 · 8)

C. 7(3 + 10)

d. 497(268 + 574)

D. (3 · 5) + (3 · 17)

e. (7 · 3) + (7 · 10)

E. 9(6 + 4)

f. (9 · 6) + (9 · 5)

F. (10 · 8) + (10 · 4)

g. 150(29 + 62)

G. (497 · 268) + (497 · 774)

You have been working with the <u>Distributive</u> <u>Property</u>.
Here are two examples:

a. $5(8 + 2) = (5 · 8) + (5 · 2)$

b. $11(9 + 8) = (11 · 9) + (11 · 8)$

In example <u>a</u>, 5(8 + 2) is easier to work. $5(8 + 2) = 5 · 10 = 50$
In example <u>b</u>, (11 · 9) + (11 · 8) is easier to work.
$$(11 · 9) + (11 · 8) = 99 + 88 = 187$$

---

4. Use the distributive property to write a second expression equal to the first. Circle the expression you think was easier to work.

a. 5(4 + 9)

b. 7(10 + 8)

c. (15 · 10) + (15 · 6)

d. (13 · 14) + (13 · 16)

Answers (cont.)

3. a. <u>D</u>   b. <u>F</u>   c. <u>A</u>   d. <u>no match</u>   e. <u>C</u>   f. <u>no match</u>   g. <u>B</u>

4. Answers will vary.

a. $5(4 + 9) = \boxed{(5 \cdot 4) + (5 \cdot 9)}$

b. $7(10 + 8) = \boxed{(7 \cdot 10) + (7 \cdot 8)}$

c. $\boxed{(15 \cdot 10) + (15 \cdot 6)} = 15(10 + 6)$

d. $(13 \cdot 14) + (13 \cdot 16) = \boxed{13(14 + 16)}$

ALGEBRAIC STRIPS AND SINGLES

1. Match the algebraic expression with the geometric figure it describes. <u>Note</u>: 3·X means 3 times X and is often written 3X.

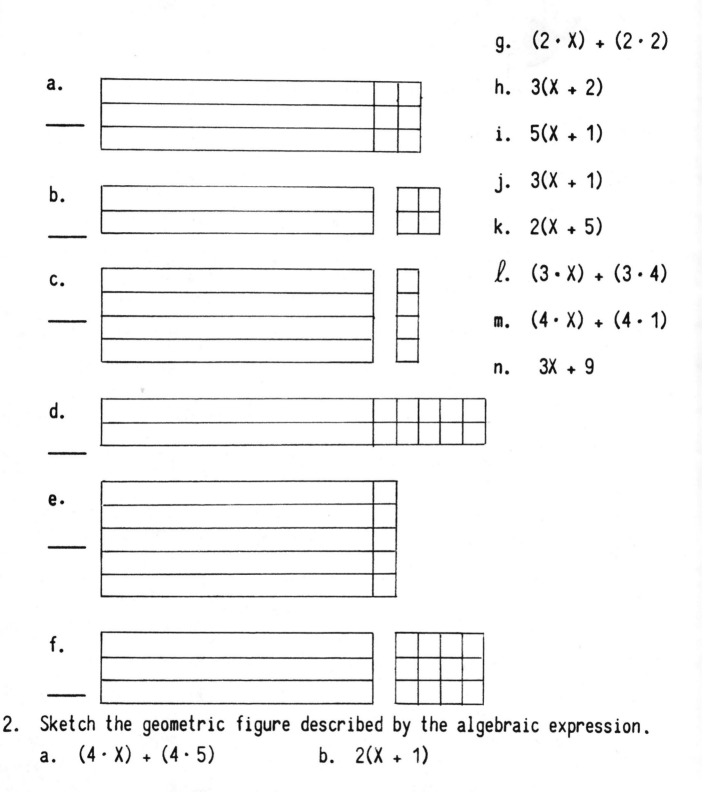

a. ___

b. ___

c. ___

d. ___

e. ___

f. ___

g. (2·X) + (2·2)

h. 3(X + 2)

i. 5(X + 1)

j. 3(X + 1)

k. 2(X + 5)

ℓ. (3·X) + (3·4)

m. (4·X) + (4·1)

n. 3X + 9

2. Sketch the geometric figure described by the algebraic expression.
   a. (4·X) + (4·5)   b. 2(X + 1)

3. Why is the lesson called <u>Algebraic</u> <u>Strips</u> <u>And</u> <u>Singles</u>? How is it different than the last lesson, <u>Strips</u> <u>And</u> <u>Singles</u>?

Algebraic <u>Strips</u> <u>And</u> <u>Singles</u>

Mathematics teaching objectives:

  . Visualize and illustrate the distributive property using geometric figures.

  . Practice using the distributive property with a variable.

Problem-solving skills pupils <u>might</u> use:

  . Make and use a drawing.

  . Look for patterns.

  . Generalize a problem solution to include other solutions.

Materials needed:

  . Cuisenaire rods (optional)

Comments and suggestions:

  . Cuisenaire rods could be used on the overhead to demonstrate problems 1 and

  . This lesson should be preceded by <u>Strips</u> <u>And</u> <u>Singles</u> (page 39). Both shoul
    be used early in the year when introducing the use of variable.

  . Assist in the interpretation of the drawings by working problems 1, 2, and
    with the class. Emphasize that the variable <u>X</u> is a symbol for <u>any</u> strip
    length one might choose. In the preceding lesson, the strip length was
    restricted to 10.

  . Problems 6 and 7 afford opportunities to discuss key mathematical ideas.
    See comments with the answers. Extend the discussion to include the most
    general expression of the distributive property:

$$a(b + c) = ab + ac \quad \text{where a, b, and c}$$
can be any of the numbers used in arithmetic.

Answers:

  1. a. <u>h</u>   b. <u>g</u>   c. <u>m</u>   d. <u>k</u>   e. <u>i</u>   f. <u>l</u>

  2. a.

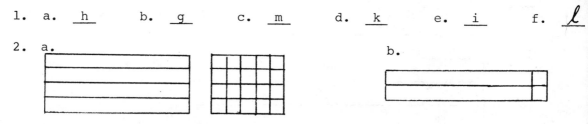

  3. This lesson uses the variable <u>X</u> rather than only numbers with a definite
     value. <u>X</u> is a symbol for <u>any</u> number one chooses to use.

  4. a. <u>D</u>   c. <u>no match</u>   e. <u>G</u>   g. <u>B</u>
     b. <u>F</u>   d. <u>A</u>   f. <u>C</u>

  5. a. <u>D</u>   c. <u>G</u>   e. <u>F</u>   g. <u>no match</u>
     b. <u>B</u>   d. <u>no match</u>   f. <u>C</u>

  6. $6(X + 2) = (X + 2) + (X + 2) + (X + 2) + (X + 2) + (X + 2) + (X + 2) =$

     $(X + X + X + X + X + X) + (2 + 2 + 2 + 2 + 2 + 2) = 6X$

     Remind pupils they are using the same rearrangement property with a varial
     as they used when adding a column of figures in arithmetic.

  7. The first example is one special case from arithmetic; the second is a
     special case of a <u>family</u> of arithmetic examples. <u>X</u> can be any number use
     in arithmetic.

4. Match the expressions needed to show examples of the
    distributive property. Not all expressions can be matched.

a.	$4(X + 5)$	A.	$X(X + 3)$
b.	$2(X + 3)$	B.	$(^-4 \cdot X) + (^-4 \cdot 2)$
c.	$(5 \cdot 2X) + (5 \cdot 4)$	C.	$(3 + X)3$
d.	$(X \cdot X) + (X \cdot 3)$	D.	$(4 \cdot X) + (4 \cdot 5)$
e.	$2X(X + 4)$	E.	$5(X + 4)$
f.	$(3 \cdot 3) + (X \cdot 3)$	F.	$(2 \cdot X) + (2 \cdot 3)$
g.	$^-4(X + 2)$	G.	$(2X \cdot X) + (2X \cdot 4)$

5. Match the expressions needed to show examples of the
    distributive property. Not all expressions can be matched.
    The expressions in the right column have been simplified.

a.	$4(X + 7)$	A.	$X^2 + Y$
b.	$X(X + 1)$	B.	$X^2 + X$
c.	$(X + 9)2$	C.	$Y^2 - 3Y$
d.	$X(X + Y)$	D.	$4X + 28$
e.	$3(2X + 5)$	E.	$10XY$
f.	$Y(Y - 3)$	F.	$6X + 15$
g.	$10(X + Y)$	G.	$2X + 18$

6. Show by using repeated addition that $6(X + 2)$ does equal $6X + 12$.

7. In what way are these examples of the distributive property
    different?
$$5(2 + 7) = 5 \cdot 2 + 5 \cdot 7$$
$$5(X + 7) = 5X + 5 \cdot 7$$

ACROSS AND DOWN
(Ideas for Teachers)

The 7th-grade problem-solving packet contains a drill-and-practice activity entitled "Across and Down." The first part of this activity is reproduced below.

When Jill came to class she found this problem on the board.

17	8	=
2	13	=

Add across.
Add down.
Add your results,
    across and down.
Put these answers in
    the triangles.
WHAT DO YOU NOTICE?

Jill was convinced that this was a very special case. She announced the teacher used special numbers to make it work. So Jill made up one of her own. Try this one to see if it works out.

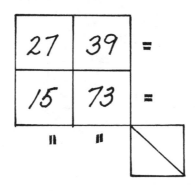

Across And Down

Mathematics teaching objectives:

- Use algebra to make explanations of numerical discoveries.
- Use the rearrangement (commutative and associative) properties of addition and multiplication.
- Practice addition, subtraction, multiplication, and division of integers and polynomials.

Problem-solving skills pupils might use:

- Look for patterns.
- Create new problems by varying a given one.
- Make explanations based upon data.

Materials needed:

- None

Comments and suggestions:

- The suggested activities can be used at various times during the year but are especially appropriate at the beginning of the year when integers and properties are introduced and later when polynomials are studied.
- The basic structure of this lesson provides many opportunities for pupils to invent and explore "what if" kinds of questions.
- Usually, several pupils will create 2 by 2 problems that "don't work." Suggest they check their work with classmates.

Answers:

1 and 2.

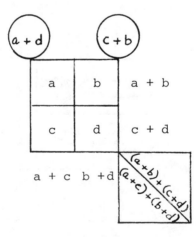

Pupils should be able to show this kind of an algebraic justification.

Across and Down (Cont.)

Variations of this activity can be used effectively in several different places in the algebra course. Here are some suggestions that are worthy of student exploration.

1. Why do problems like these work out the way they do?

2. What if we put "ears" on the figure, add diagonally, and then find the sum of the "ears?"

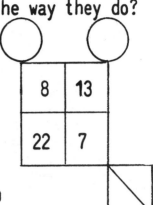

3. What if we use a 3 by 3, rather than a 2 by 2? Will the "ears" property still work? Under what conditions will the "ears" property work?

4. What if decimals or negative numbers are used?

5. What if we multiply the numbers (and the results) rather than add?
What if subtraction is used?
What if division is used?

6. What if we use polynomials in the boxes as shown at the right?

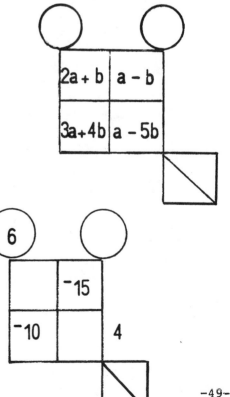

7. What if we put numbers or polynomials in other parts of the figure? Do we always have to use at least four numbers to complete the problem? How many numbers are needed to complete a problem in a 3 by 3 arrangement?

Answers:

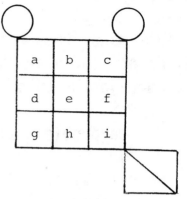

3. In most 3 by 3 examples pupils
    create, the "ears property" will
    not work. However, the "ears
    property" will work if
    $e = b + d + f + h$.

4. Decimals and negative numbers
    also work.

5. Multiplication will also work and can
    be justified by the commutative and
    associative properties. Even though
    subtraction and division do not have these
    properties, surprisingly, when subtraction
    or division are used, the "magic sum" is
    still the same.

Subtraction

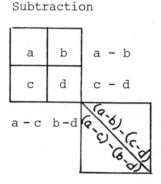

Division

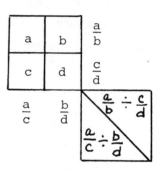

6. See the answers to problems 1, 2, and 5.

7. In general, in a 2 by 2, four numbers appropriately placed are needed to
    complete the problem. (It is possible to use only three numbers if one
    of them is in the answer box.) In general, in a 3 by 3, one needs to
    have nine numbers appropriately placed.

PARENTHESES

1. Find the answers. Remember, do the operation within the parentheses first. Otherwise, work from left to right.

   a. $8 + (^-5 + 6)$

   b. $8 + ^-5 + 6$

   c. $^-3 + (^-4 + ^-5)$

   d. $^-3 + ^-4 + ^-5$

   e. $12 - (^-8 + 9)$

   f. $12 - ^-8 + 9$

   g. $^-10 + (5 + ^-2)$

   h. $^-10 + 5 + ^-2$

   i. $7 + (^-15 - ^-5)$

   j. $7 + ^-15 - ^-5$

   k. $^-6 + (^-2 + ^-4)$

   ℓ. $^-6 + ^-2 + ^-4$

   m. $^-1 + (4 - ^-2)$

   n. $^-1 + 4 - ^-2$

   o. $3 + (^-8 + 5)$

   p. $3 + ^-8 + 5$

2. The exercises in problem 1 come in pairs. In all cases but one, the two answers are the same.

   a. How are the expressions in each pair alike? How different?

   b. Which pair does not have the same answer?

   c. What makes this pair different from the others?

<u>Parentheses</u>

Mathematics teaching objectives:

- Practice addition and subtraction of integers.
- Discover the procedures for correctly removing parentheses in numerical expressions.
- Emphasize number properties (associative property of addition).

Problem-solving skills pupils <u>might</u> use:

- Look for patterns.
- Find likenesses and differences and make comparisons.
- Make predictions based upon data.

Materials needed:

- None

Comments and suggestions:

- This lesson can be used after addition and subtraction with integers have been studied. It also sets the stage for further work with parentheses, e.g. $2x - (x + 3)$. The main purpose is for pupils to recognize that a minus sign directly in front of parentheses "does make a difference." More practice with this concept will be needed. In other words, the associative property is true for addition but not for subtraction. Adjustments need to be made if an analogous property is to work for subtraction.
- Do this lesson in class. Have pupils work the exercises and answer the questions in problems 1 and 2 on their own. Encourage them to discuss their responses to problem 2 with a classmate. The rest of the lesson should be done independently. Offer assistance where necessary but be as nondirective as you can manage.
- Observe pupils' responses to problem 3. They likely will overgeneralize failing to note that an adjustment was made in exercise 3j. Mention this overgeneralizing tendency when the lesson is summarized.

Answers:

See page 54.

Parentheses (cont.)

3. The following exercises come in pairs. Without working them, predict which pairs will <u>NOT</u> have the same answer. Three pairs have different answers.

a. $3 - (^-5 + 4)$

g. $4 + (2 + ^-6)$

b. $3 - ^-5 + 4$

h. $4 + 2 + ^-6$

c. $^-2 + (5 + ^-6)$

i. $^-5 - (^-6 - 8)$

d. $^-2 + 5 + ^-6$

j. $^-5 - ^-6 + 8$

e. $10 - (^-1 - ^-3)$

k. $^-10 - (3 + ^-4)$

f. $10 - ^-1 - ^-3$

$\ell.$ $^-10 - 3 + ^-4$

4. Now, find the answer to each exercise in problem 3. Do your answers agree with your predictions? If not, explain why.

5. Which problem below is done correctly? How can you tell without working the problems?

a. $5 - (3 - ^-2) = 5 - 3 - ^-2$

b. $5 - (3 - ^-2) = 5 - 3 + ^-2$

6. Debbie was discussing some "parentheses" problems with Sandy.

"Watch out when a minus sign comes before the parentheses."

What did Debbie mean by this? Make up an example to help you answer this question.

<u>Parentheses</u>

Answers:

1.
a.	9		i.	⁻3	
b.	9		j.	⁻3	
c.	⁻12		k.	⁻12	
d.	⁻12		*l.*	⁻12	
e.	11		m.	5	
f.	29		n.	5	
g.	⁻7		o.	0	
h.	⁻7		p.	0	

2. a. Same numbers are used in each expression and in the same left-to-right order. One of the expressions in each pair does not have parentheses.

    b. e and f

    c. In <u>e</u>, a subtraction sign precedes the parentheses. In the others, an addition sign precedes the parentheses.

3. Answers will vary. a and b, e and f, k and *l* have the same answers.

4.
a.	4		g.	0
b.	12		h.	0
c.	⁻3		i.	9
d.	⁻3		j.	9
e.	8		k.	⁻9
f.	14		*l.*	⁻17

5. b is done correctly. Pupils will discover that if a minus sign immediately precedes a set of parentheses, a true statement results if the parentheses are removed and the operation sign within the parentheses is changed. This "change" is mathematically equivalent to the usual rule of <u>add</u> <u>the</u> <u>opposite</u> found in the textbooks.

6. See the problem and answer for number 5.

PERIMETER

Perimeter is the distance around. Study these examples.

      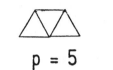   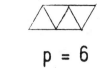

p = 3   p = 4   p = 5   p = 6

1. Complete the tables below.

a.

...

Number of Triangles	1	2	3	4	5	6	10			n
Perimeter	3	4	5					40		

b.

Number of Squares	1	2	3	4	5	6	10		n
Perimeter	4	6						62	

c.

Number of Pentagons	1	2	3	4	5	6	10		n
Perimeter	5	8						92	

d.

Number of Hexagons	1	2	3	4	5	6	10		n
Perimeter	6							102	

e. Suppose the figures have 20 sides and are lined up in the same way.

Number of 20-sided figures	1	2	3	4	5	6	10		n
Perimeter	20							362	

Perimeter

Mathematics teaching objectives:

- Review the concept of perimeter.
- Write an algebraic expression for the nth case of a linear relation.

Problem-solving skills pupils might use:

- Look for patterns.
- Make and use a drawing.
- Make generalizations based upon data.

Materials needed:

- None

Comments and suggestions:

- This lesson should come early in the year when pupils are having their first experience with algebraic expressions.
- Much of this lesson should be done during class. Provide units when needed.
- Pupils will notice patterns within the charts. Some will notice patterns from one chart to another, especially in problem number 1.
- The general expressions may be difficult to find. But once the first few have been found, pupils should be encouraged to look back for clues that may help them discover the algebraic expression for the nth case.
- Strongly encourage pupils to give some time and effort on problem 4. This is their chance to do "their own thing."

Answers:

1. a.

Number of Triangles	1	2	3	4	5	6	10	38	n
Perimeter	3	4	5	6	7	8	12	40	$n + 2$

b.

Number of Squares	1	2	3	4	5	6	10	30	n
Perimeter	4	6	8	10	12	14	22	62	$2n + n$

c.

Number of Pentagons	1	2	3	4	5	6	10	30	n
Perimeter	5	8	11	14	17	20	32	92	$3n + 2$

d.

Number of Hexagons	1	2	3	4	5	6	10	25	n
Perimeter	6	10	14	18	22	26	42	102	$4n + 2$

e.

Number of 20-sided Figures	1	2	3	4	5	6	10	20	n
Perimeter	20	38	56	74	92	110	182	362	$18n + 2$

Perimeter (cont.)

2. In this problem, the figures look like .

    a.

Number of Figures	1	2	3	4	5	10		n
Perimeter	8	14					92	

    b. In what way are the "fittings" here different from those in problem 1?

    c.

Number of Figures	1	2	3	4	5	10		n
Perimeter	8						104	

3. In this problem, the figures look like .

    a.

Number of Figures	1	2	3	4	5	10		n
Perimeter	10	18					162	

    b.

Number of Figures	1	2	3	4	5	10		n
Perimeter	10						148	

4. Use figures like this:

    a. Create your own "fittings."

    b. Use your design. If the number of figures is n, what is the perimeter?

Perimeter

Answers:

2. a.

Number of Figures	1	2	3	4	5	10	15	n
Perimeter	8	14	20	26	32	62	92	$6n + 2$

b. The shared boundary is not a complete side for both figures. The shared boundary has a measure of 1.

c.

Number of Figures	1	2	3	4	5	10	25	n
Perimeter	8	12	16	20	24	44	104	$4n + 4$

Notice the shared boundary has a measure of 2.

3. a.

Number of Figures	1	2	3	4	5	10	40	n
Perimeter	10	18	26	34	42	82	162	$8n + 2$

b.

Number of Figures	1	2	3	4	5	10	24	n
Perimeter	10	16	22	28	34	64	148	$6n + 4$

4. Answers will vary.

SEMESTER EXAM

Claude didn't do very well on the semester exam in algebra. In fact, he had only 3 correct.

Semester Exam   *Claude*

Simplify, if possible.

1. $3a + 2a$ $5a^2$

2. $4n - 5n$ $-n$

3. $2b^2 + 4b^2$ $6b^4$

4. $2a \cdot 3b$ $5ab$

5. $2c - (c - b)$ $c - b$

6. $(2rs)^2$ $4r^2s^2$

7. $2a + b + 3a + 4b$ $6a^2 + 4b^2$

8. $\dfrac{2a^6}{a^3}$ $2a^2$

9. $\dfrac{d + 3}{d}$ 3

10. $4 - (a + 7)$ $^-3 - a$

---

1. Study Claude's paper. Use your intuition. Decide which of his answers should be marked wrong.

2. At this time, you may not know how to simplify some of the expressions on the Semester Exam. (Evidently Claude didn't either.) But at least you can determine which are correct by substituting numbers. In each problem, choose a value for the variables. Evaluate each expression. Also evaluate Claude's answer. Clearly indicate the numbers you used.

    a. Which ones did Claude get correct?

    b. How do these results compare with your original "hunches?"

Semester Exam

Mathematics teaching objectives:

. Substitute values to evaluate expressions.

Problem-solving skills pupils <u>might</u> use:

. Look for patterns and/or properties.

. Guess and check.

Materials needed:

. None

Comments and suggestions:

. This lesson should be done early in the school year and, or course, prior to any "formal" work with simplification. The lesson is short--probably less than 20 minutes.

. The main purpose of the lesson is for pupils to check algebraic simplifications by substituting in values for the variables and then working the resulting exercise.

. Suggest that numbers other than 0 and 1 be used as values for the variable. Illustrate this by using Claude's test and substituting 0 for <u>a</u> in exercise 8, 0 for <u>d</u> in exercise 9, and 1 for <u>b</u> in exercise 3.

. Upon completing the lesson, pupils should be more cautious in using their intuition in simplifying algebraic expressions.

Answers:

1. Claude missed 1, 3, 4, 5, 7, 8, and 9. The correct answers are:

      1. $5a$

      3. $6b^2$ <u>Note</u>:   Pupils were not expected to find correct answers.

      4. $6ab$

      5. $c + b$

      7. $5a + 5b$

      8. $2a^3$

      9. $\dfrac{d + 3}{3}$ (cannot be simplified)

2. a. Claude's answers to problems 2, 6, and 10 were correct.

    b. Answers will vary.

LIKE TERMS

In each case, determine which answer is correct. Substitute numbers to help you decide.

1. $3g + 2h =$
   a. $6gh$
   b. $5gh$
   c. $6g^2h^2$
   d. None of these

2. $3n + 2n =$
   a. $5n^2$
   b. $6n^2$
   c. $5n$
   d. None of these

3. $5x + x =$
   a. $5x^2$
   b. $6x^2$
   c. $6x$
   d. None of these

4. $2p + 3p + 7p =$
   a. $12p^3$
   b. $42p^3$
   c. $12p$
   d. None of these

5. $4p + 5q =$
   a. $20\ pq$
   b. $9pq$
   c. $20p^2q^2$
   d. None of these

6. $8n - 2n =$
   a. $6n^2$
   b. $6n$
   c. 4
   d. None of these

7. $2ef + 3ef =$
   a. $5e^2f^2$
   b. $6e^2f^2$
   c. $5ef$
   d. None of these

8. $3r^2 + 4s^2 =$
   a. $7r^2s^2$
   b. $12r^2s^2$
   c. $7(r + s)^2$
   d. None of these

9. $2x^2 + 5x^2 =$
   a. $7x^4$
   b. $10x^4$
   c. $7x^2$
   d. None of these

Simplify if possible.

10. $4a + 6a$

11. $3n + 9n$

12. $11x - 7x$

13. $5p + 9r$

14. $5b + 6b - 3b$

15. $2xy + 8xy$

16. $2a^2 + 3b^2$

17. $5a - 8a$

18. $a + b + c$

19. $^-n - n - n$

Like Terms

Mathematics teaching objectives:

- Add like terms.

- Substitute numbers to evaluate expressions.

Problem-solving skills pupils <u>might</u> use:

- Guess and check.

- Make necessary computations needed for the solution.

- Study the solution process.

Materials needed:

- None

Comments and suggestions:

- This lesson is intended as an introduction to combining similar or like terms. More practice will be needed.

- Encourage substitution of small numbers for each variable as a method for finding an equivalent expression for the answer, but avoid using 0 and 1 because of their special properties.

- Pupils probably will find the lesson rather easy. Plan for something else or give this toward the end of the period.

Answers:

1. d	2. c	3. c
4. c	5. d	6. b
7. c	8. d	9. c

10. 6a	15. 10xy
11. 12n	16. $2a^2 + 3b^2$
12. 4x	17. ^-3a
13. 5p + 9r	18. a + b + c
14. 8b	19. ^-3n

FIBONACCI

Leonardo Fibonacci was a leading European mathematician of the Middle Ages. He became interested in patterns similar to the ones below. What is the next term in each sequence? Can you discover how these patterns were created?

$$1, 1, 2, 3, 5, 8, 13, 21, \ ?$$
$$2, 5, 7, 12, 19, 31, 50, 81, \ ?$$
$$10, 3, 13, 16, 29, 45, 74, 119 \ ?$$

Use the same rule to complete the patterns below.

1. 3, 7, 10, ___, ___, ___, ___

2. 3, 1, 4, ___, ___, ___, ___

3. 2, ___, 8, ___, ___, ___, ___

4. ___, 5, ___, 11, ___, ___, ___

5. 6, ___, ___, 8, ___, ___, ___

6. 3, ___ ___, 7, ___, ___, ___

7. 5, ___, ___, 5, ___, ___, ___

8. 3, ___, ___, 27, ___, ___, ___

9. 1, ___, ___, ___, 14, ___, ___

10. 3, ___, ___, ___, 30, ___, ___

11. 4, ___, ___, ___, ___, 27, ___, ___

12. 7, ___, ___, ___, ___, ___, 51, ___, ___

Fibonacci

Mathematics teaching objectives:

. Use mental computation to add whole numbers and fractions.

. Use algebraic equations to solve Fibonacci sequence problems.

Problem-solving skills pupils might use:

. Look for patterns.

. Guess and check.

. Use mathematical symbols to describe situations.

Materials needed:

. None

Comments and suggestions:

. Elicit responses from the class as you work through the first few problems with them and until they seem to understand how the pattern is generated.

. The missing numbers in problem 13 are fractions and are difficult to find by a guess and check strategy. This difficulty serves as motivation for finding a more efficient procedure. Problem 14 shows the power of algebra.

Answers:

1. 3, 7, 10, 17, 27, 44, 71

2. 3, 1, 4, 5, 9, 14, 23

3. 2, 6, 8, 14, 22, 36, 58

4. 1, 5, 6, 11, 17, 28, 45

5. 6, 1, 7, 8, 15, 23, 38

6. 3, 2, 5, 7, 12, 19, 31

7. 5, 0, 5, 5, 10, 15, 25

8. 3, 12, 15, 27, 42, 69, 111

9. 1, 4, 5, 9, 14, 23, 37, 60

10. 3, 8, 11, 19, 30, 49, 79

11. 4, 3, 7, 10, 17, 27, 44, 71

12. 7, 2, 9, 11, 20, 31, 51, 82, 133

13. Try to complete the
    sequence at the right.   8, ___, ___, ___, 27
    Why is this problem
    more difficult to solve than the others?

14. Algebra can be used to solve problem 13. Study the steps
    below. Then solve the problem.

$$8, \underline{\quad\quad}, \underline{\quad\quad}, \underline{\quad\quad}, 27$$

$$8, \underline{\;a\;}, \underline{(8 + a)}, \underline{(8 + 2a)}, 27$$

$$\underset{\substack{\text{3rd} \\ \text{term}}}{(8 + a)} + \underset{\substack{\text{4th} \\ \text{term}}}{(8 + 2a)} = \underset{\substack{\text{5th} \\ \text{term}}}{27}$$

15. 5, ___, ___, ___, 23

16. 10, ___, ___, ___, 40

17. 3, ___, ___, ___, ___, 50

18. 5, ___, ___, ___, ___, ___, 60

19. 10, ___, ___, ___, ___, 0

20. Make up a Fibonacci problem of your own. Leave some blank
    spaces. First solve it yourself. Then give it to someone
    else to solve.

Answers:

13. 8, $3\frac{2}{3}$, $11\frac{2}{3}$, $15\frac{1}{3}$, 27

14. See problem 13.

15. 5, a, 5 + a, 5 + 2a, 23

   5, $4\frac{1}{3}$, $9\frac{1}{3}$, $13\frac{2}{3}$, 23

16. 10, a, 10 + a, 10 + 2a, 40

   10, $6\frac{2}{3}$, $16\frac{2}{3}$, $23\frac{1}{3}$, 40

17. 3, a, 3 + a, 3 + 2a, 6 + 3a, 50

   3, 8.2, 11.2, 19.4, 30.6, 50

18. 5, a, 5 + a, 5 + 2a, 10 + 3a, 15 + 5a, 60

   5, $4\frac{3}{8}$, $9\frac{3}{8}$, $13\frac{6}{8}$, $23\frac{1}{8}$, $36\frac{7}{8}$, 60

19. 10, a, 10 + a, 10 + 2a, 20 + 3a, 0

   10, $^-6$, 4, 2, $^-2$, 0

20. Problems will vary.

EVALUATING EXPRESSIONS

Use the clues in the table. Complete each row. For the last part, create an expression that gives the answers shown.

	x	y	$x + y$	xy	$x - y$	$\dfrac{x}{y}$	$2x + y$	$4x - 3y$	$x^2 - y^2$	
1.	6	3								⁻3
2.	4		6							⁻2
3.				12	⁻1					1
4.			11	24						⁻5
5.		4							9	⁻1
6.	8						18			⁻6
7.					9	4				⁻9
8.		⁻5						15		⁻5

-67-

© PSM 82

Evaluating Expressions

Mathematics teaching objectives:

- Substitute numbers to evaluate expressions.
- Find an algebraic expression which must conform to certain conditions.
- Practice solving equations.

Problem-solving skills pupils _might_ use:

- Work backwards.
- Guess and check.
- Make decisions based upon data.
- Make a systematic list.

Materials needed:

- None

Comments and suggestions:

- This lesson should be used after pupils have had some background in evaluating expressions but can be used before a formal unit on equation solving.
- Explain that x and y have the same values throughout problem 1, but then change for problem 2, etc.
- Pupils may need to verbalize what the problem is saying. For example, in problem 7, the pupil is looking for two numbers whose difference is 9 and whose quotient is 3.
- Problem 4 appears to have two sets of answers--8,3 and 3,8. The expression found for the $\frac{x}{y}$ column and for the last column eliminates 3,8.

Answers:

	x	y	x + y	xy	x - y	$\frac{x}{y}$	2x + y	4x - 3y	$x^2 - y^2$	y - x
1.	6	3	9	18	3	2	15	15	27	-3
2.	4	2	6	8	2	2	10	10	12	-2
3.	3	4	7	12	⁻1	$\frac{3}{4}$	10	0	⁻7	1
4.	8	3	11	24	5	$2\frac{2}{3}$	19	23	55	⁻5
5.	5	4	9	20	1	$1\frac{1}{4}$	14	8	9	⁻1
6.	8	2	10	16	6	4	18	26	60	⁻6
7.	12	3	15	36	9	4	27	39	135	⁻9
8.	0	⁻5	⁻5	0	5	0	⁻5	15	⁻25	⁻5

MATH-MAGIC

Felix, the class trickster, is at it again. Here's the problem he gave the class.

"First, I'm going to place some coins in Julie's hand.
Now, each of you must follow these directions carefully --

. Think of a number between 1 and 10.
. Add your number to one more than your number.
. Add 9.
. Divide by 2.
. Subtract your original number.
. Divide by 2.

Now, I hope all of your work has been done very carefully.
If so, your final answer should be the number of coins in Julie's hand."

"No way," said Lori.
"You must be joking," exclaimed Bryce.
"Impossible--Ridiculous--Absurd," shouted the class.

So, Felix went over the directions once again. And the comments from the class were the same.

"Felix has finally goofed. It's about time he made a mistake!"

Felix then had Julie show the class the coins she had. Felix had NOT made a mistake. The trickster had used a penny that had been cut in half.

Class Exercises

1. Why was this Felix's last magic trick for a while?

2. Try the same trick by using a number between 10 and 20.
   Does it still work?

Math-Magic

Mathematics teaching objectives:

. Use variables to explain number tricks.

. Practice in writing and solving equations.

Problem-solving skills pupils _might_ use:

. Look for patterns.

. Make predictions based upon data.

. Make explanations based upon data.

Materials needed:

. $2\frac{1}{2}$ coins (optional)

Comments and suggestions:

. Use early in the year when pupils are using variables in writing expressions and making simplifications.

. The opening 'trick' is most effective if actual coins can be used, but you should practice the trick before using it in class!

. Most of the lesson should be done during class so the teacher can provide assistance in making the algebraic explanations.

Answers:

Class Exercises

1. Who knows! It could have been that the teacher was tired of the trickster's tricks. Or, since it is illegal to deface a coin, perhaps the trickster got caught.

2 and 3. The trick will work for any number.

4. a. Yes, it still works.

    b. Let N   represent the number
            3N   result of multiplying by 3
            3N + 8   result of adding 8
            4N + 8   result of adding the original number
            N + 2   result of dividing by 4
            N + 1   result of subtracting 1

Exercises

See page 74.

3. Do you think the trick will work if you start with a number greater than 20? Try it.

    Here's why the "trick" works. Notice how the variable is used.

Let N	represent the number.
N + 1	represents one more than the number.
2N + 1	is the sum of the two numbers.
2N + 10	result of adding 9
N + 5	result of dividing by 2
5	result of subtracting the original number
$2\frac{1}{2}$	result of dividing by 2

4. Here's another number trick.

    Pick a number between 10 and 20.
    Multiply by 3.
    Add 8.
    Add your original number.
    Divide by 4.
    Subtract 1.

    You should end up with 1 more than your original number.

    a. Do the same trick, but begin with a number greater than 20. Does it work?

    b. Use variables to show why this trick works.

Math-Magic (cont.)

Exercises

First work through the following "math-magic" problems using numbers. Then use a variable to show why the magic works.

1. Pick a number.
   Add 5.
   Multiply by 3.
   Subtract 15.
   Divide by 3.

   You should end up with your original number. Remember, after you've tried some numbers, use a variable to show why the trick works.

2. Pick a number.
   Multiply by 12.
   Add the number of inches in a yard.
   Subtract the number of eggs in a dozen.
   Divide by the number of inches in a foot.
   Subtract your original number.
   Your answer should be 2.

3. Pick a number between 1 and 10.
   Multiply by 99.
   Add your original number.
   Divide by 100.
   Your answer should be the same as your original number.

4. Write down the number of the month in which you were born.
   Multiply by 2.
   Add 5.
   Multiply by 50.
   Add your age.
   Subtract 365.
   Add 115.

   Your age will be the last 2 digits. The first digits will be the month in which you were born.

5. Create a "math-magic" problem of your own.

Exercises

1. N

    N + 5   add 5

    3N + 15   multiply by 3

    3N   subtract 15

    N   divide by 3

2. N

    12N   multiply by 12

    12N + 36   add 36

    12N + 24   subtract 12

    N + 2   divide by 12

    2   subtract original number

3. N

    99N   multiply by 99

    100N   add original number

    N   divide by 100

4. m   month born

    2m   multiply by 2

    2m + 5   add 5

    100m + 250   multiply by 50

    100m + 250 + A   add age

    100m - 115 + A   subtract 365

    100m + A   add 115

5. Problems will vary.

FRACTIONS TEST

Claude's answers to this True-False test are shown. Use what you know about arithmetic fractions to grade the test. Circle the ones he missed.

T   1. $\dfrac{N}{3} = \dfrac{10N}{30}$

T   2. $\dfrac{\cancel{N} + 3}{\cancel{N}} = 4$

T   3. $\dfrac{N}{2} \cdot \dfrac{N}{3} = \dfrac{N^2}{6}$

F   4. $\dfrac{2N}{5} - \dfrac{N}{3} = \dfrac{N}{2}$

T   5. $\dfrac{3}{\cancel{N}} + \dfrac{\cancel{N}}{3} = 2$

F   6. $\dfrac{N}{10} + \dfrac{N}{10} = \dfrac{2N}{20}$

F   7. $N - \dfrac{N}{3} = \dfrac{2N}{3}$

F   8. $\dfrac{(N + 1)(\cancel{N + 2})}{(\cancel{N + 2})} = N + 1$

T   9. $\dfrac{2N}{3} \cdot \dfrac{3}{2N} = 1$

F   10. $\dfrac{2N + 4}{N + 2} = 2$

Fractions <u>Test</u>

Mathematics teaching objectives:

. Extend knowledge of fractions to algebraic fractions.

. Substitute values for a variable to check an equation.

Problem-solving skills pupils <u>might</u> use:

. Recall related information.

. Solve an easier but related problem.

. Generalize a problem solution to include other solutions.

Materials needed:

. None

Comments and suggestions:

. This lesson is intended as an introduction to algebraic fractions.

. Remind the class that the variable <u>N</u> is to be treated as a number--<u>any</u> number. The rules for arithmetic fractions also apply to algebraic fractions. This lesson should be a check on the understandings pupils have of arithmetic fractions and their working knowledge of algebra in generalizing to situations involving a variable.

. Encourage substitution of numbers to evaluate both sides of the expressions. When pupils substitute numbers, it is best if they avoid the use of <u>0</u> and <u>1</u> as replacements. Show why by using zero for <u>N</u> in exercise 2 and <u>1</u> for N in exercise 10.

Answers:

Clause missed the following problems: 2, 5, 7, 8, 10.

No. 2 is false. It cannot be reduced.

No. 5 is false. "Canceling" is not done between terms in an addition problem.

No. 7 is true. Pupils often do not relate N to $\frac{3N}{3}$ or $\frac{N}{3}$ to $\frac{1N}{3}$.

No. 8 is true. This reduction is correct.

No. 10 is true. Substituting various values for N shows this always to be true except for N = $^-2$.

CATCH 22

1. a. Choose two 1-digit numbers. ___ ___
    b. Use the numbers to make two 2-digit numbers. ___ ___
    c. Add the two 2-digit numbers. _____
    d. Add the two numbers you chose. _____
    e. Divide the answer to (c) by the answer to (d). _____
    f. Compare your result with your classmates.

2. a. Choose three 1-digit numbers. ___ ___ ___
    b. Use these numbers to make six different 2-digit numbers.

       ___ ___ ___ ___ ___ ___
    c. Add the six numbers. _____
    d. Add the three numbers you chose. _____
    e. Divide the answer to (c) by the answer to (d). _____
    f. Compare your result with your classmates.

3. Repeat the investigation with three different 1-digit numbers.
    Show your steps.

4. a. Suppose you chose four 1-digit numbers and performed the
       same type of investigation. What answer do you think you
       would get? _____
    b. Do it here. Choose four 1-digit numbers. ___ ___ ___ ___
    c. Use the numbers to make twelve different 2-digit numbers.

       ___ ___ ___ ___ ___ ___ ___ ___ ___ ___ ___ ___
    d. Complete the investigation.

5. Use variables to show why the investigation in 1 and 2 works.

6. Predict the "magic" number for an investigation using eight
    1-digit numbers. _____

Catch 22

Mathematics teaching objectives:

. Practice computation skills.

. Use variables to make an explanation of a number "trick."

Problem-solving skills pupils <u>might</u> use:

. Look for a pattern.

. Make a systematic list.

. Make an algebraic explanation.

. Make predictions based upon data.

Materials needed:

. None

Comments and suggestions:

. This activity also works well without using an algebraic explanation. Lower-ability pupils can find the pattern and predict the result in problem 6.

. Some pupils will need help making the systematic list so all the different two-digit numbers are found in problems 2, 3, and 4.

. Remind the class that a number such as 28 can be written as $2 \cdot 10 + 8$. Have pupils generalize this expression when the digits are two variables, a and b.

Answers:

1. e. 11   2. e. 22   3. e. 22   4. d. 33

5. Using the three digits a, b, c

$$10a + b$$
$$10a + c$$
$$10b + a$$
$$10b + c$$
$$10c + a$$
$$\underline{10c + b}$$
$$22a + 22b + 22c$$

and $\dfrac{22a + 22b + 22c}{a + b + c} = \dfrac{22(a + b + c)}{a + b + c} = 22$

6. 77

EVALUATION TIME

Complete each table.

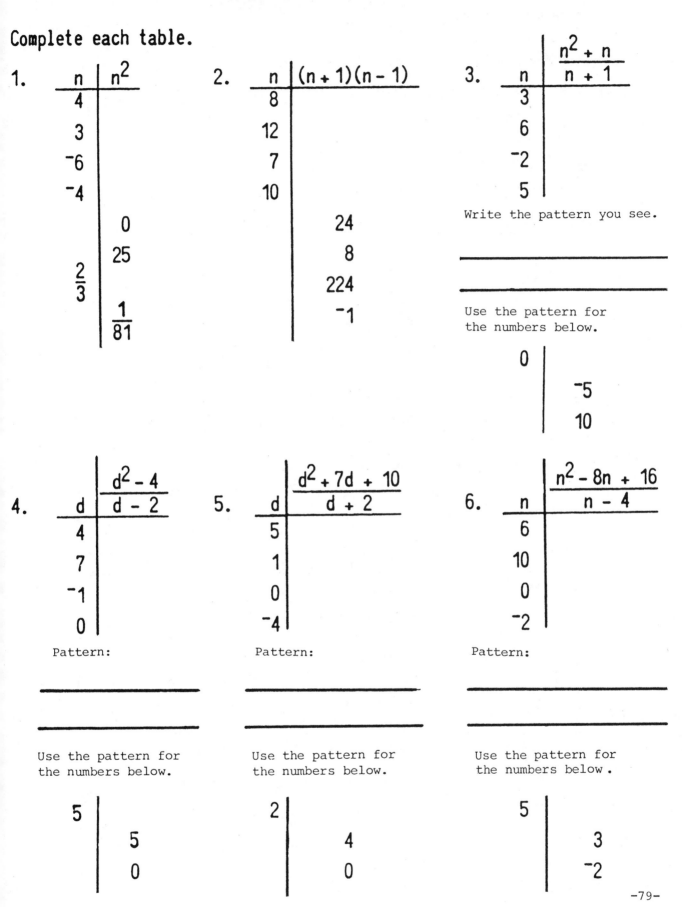

1.

n	n^2
4	
3	
⁻6	
⁻4	
	0
	25
$\frac{2}{3}$	
	$\frac{1}{81}$

2.

n	$(n+1)(n-1)$
8	
12	
7	
10	
	24
	8
	224
	⁻1

3.

n	$\dfrac{n^2 + n}{n + 1}$
3	
6	
⁻2	
5	

Write the pattern you see.

_____

_____

Use the pattern for
the numbers below.

0	
	⁻5
	10

4.

d	$\dfrac{d^2 - 4}{d - 2}$
4	
7	
⁻1	
0	

Pattern:

_____

_____

Use the pattern for
the numbers below.

5	
	5
	0

5.

d	$\dfrac{d^2 + 7d + 10}{d + 2}$
5	
1	
0	
⁻4	

Pattern:

_____

_____

Use the pattern for
the numbers below.

2	
	4
	0

6.

n	$\dfrac{n^2 - 8n + 16}{n - 4}$
6	
10	
0	
⁻2	

Pattern:

_____

_____

Use the pattern for
the numbers below.

5	
	3
	⁻2

-79-

Evaluation Time

Mathematics teaching objectives:

   . Evaluate expressions.

   . Simplify algebraic fractions.

Problem-solving skills pupils _might_ use:

   . Look for a pattern.

   . Make predictions based upon data.

   . Simplify the problem.

   . Work backwards.

Materials needed:

   . None

Comments and suggestions:

   . The activity motivates the need to simplify algebraic fractions. Pupils will need factoring skills to complete problems 9 and 10.

   . Problems 1 - 8 could be used early in the year if practice in evaluating expressions is the objective of your math lesson.

   . Problem 7 is not a mistake. It is purposely included to emphasize that $X^2 + 1$ is a prime expression and does not equal $(X + 1)(X + 1)$.

   . You may want to spend time determining the values that cannot be used-- those values that would make the denominator equal to zero. For 3, $n \neq {}^-1$. For 4, $d \neq 2$. For 5, $d \neq {}^-2$. For 6, $n \neq 4$. For 7, $X \neq {}^-1$. For 8, $e \neq 0$ or $^-15$.

Answers:

1. n	n^2
4	16
3	9
$^-6$	36
$^-4$	16
0	0
5	25
$\frac{2}{3}$	$\frac{4}{9}$
$\frac{1}{9}$	$\frac{1}{81}$

2. n	$(n + 1)(n - 1)$
8	63
12	143
7	48
10	99
5	24
3	8
15	224
0	$^-1$

3. n	$\dfrac{n^2 + n}{n + 1}$
3	3
6	6
$^-2$	$^-2$
5	5
0	0
$^-5$	$^-5$
10	10

Evaluation Time (cont.)

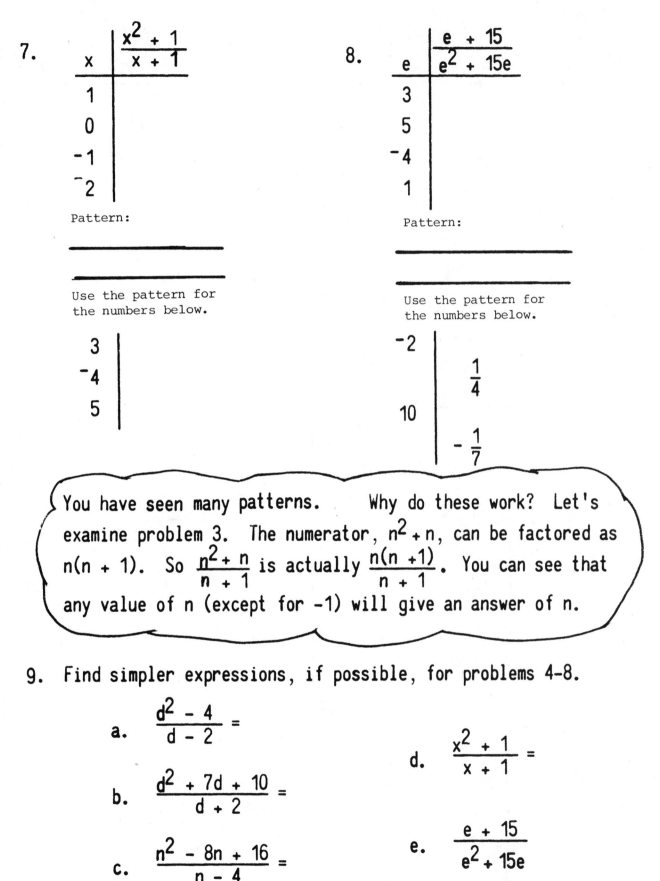

7.

x	$\dfrac{x^2 + 1}{x + 1}$
1	
0	
-1	
⁻2	

Pattern:

_____

_____

Use the pattern for the numbers below.

3	
⁻4	
5	

8.

e	$\dfrac{e + 15}{e^2 + 15e}$
3	
5	
⁻4	
1	

Pattern:

_____

_____

Use the pattern for the numbers below.

⁻2	
10	$\dfrac{1}{4}$
	$-\dfrac{1}{7}$

You have seen many patterns. Why do these work? Let's examine problem 3. The numerator, $n^2 + n$, can be factored as $n(n + 1)$. So $\dfrac{n^2 + n}{n + 1}$ is actually $\dfrac{n(n + 1)}{n + 1}$. You can see that any value of n (except for -1) will give an answer of n.

9. Find simpler expressions, if possible, for problems 4-8.

a. $\dfrac{d^2 - 4}{d - 2} =$

b. $\dfrac{d^2 + 7d + 10}{d + 2} =$

c. $\dfrac{n^2 - 8n + 16}{n - 4} =$

d. $\dfrac{x^2 + 1}{x + 1} =$

e. $\dfrac{e + 15}{e^2 + 15e}$

Answers:

4.

d	$\dfrac{d^2 - 4}{d - 2}$
4	6
7	9
⁻1	1
0	2
5	7
3	5
⁻2	0

$$\frac{d^2 - 4}{d - 2}$$

is 2 more than d

5.

d	$\dfrac{d^2 + 7d + 10}{d + 2}$
5	10
1	6
0	5
⁻4	1
2	7
⁻1	4
⁻5	0

$$\frac{d^2 + 7d + 10}{d + 2}$$

is d + 5

6.

n	$\dfrac{n^2 - 8n + 16}{n - 4}$
6	2
10	6
0	⁻4
⁻2	⁻6
5	1
7	3
2	⁻2

$$\frac{n^2 - 8n + 16}{n - 4} \text{ is } n - 4$$

7.

x	$\dfrac{x^2 + 1}{x + 1}$
1	1
0	1
⁻1	not possible
⁻2	⁻5
3	$2\frac{1}{2}$
⁻4	$⁻5\frac{2}{3}$
5	$4\frac{1}{3}$

No obvious pattern.

8.

e	$\dfrac{e + 15}{e^2 + 15e}$
3	$\frac{1}{3}$
5	$\frac{1}{5}$
⁻4	$⁻\frac{1}{4}$
1	1
⁻2	$⁻\frac{1}{2}$
4	$\frac{1}{4}$
10	$\frac{1}{10}$
⁻7	$⁻\frac{1}{7}$

$$\frac{e + 15}{e^2 + 15e} = \frac{1}{e}$$

9. a. d + 2

b. d + 5

c. n - 4

d. prime

e. $\frac{1}{e}$

Algebra

III. ALGEBRAIC EXPLANATIONS

III.   ALGEBRAIC EXPLANATIONS

Often pupils feel algebra is an isolated topic. They might know algebra
necessary for college or other mathematics and science courses, but they
 to see its real power. In these lessons, algebra is used to provide
ight as to why certain number puzzles work. If pupils have used the
blem-solving packets in previous grades, they already will be somewhat
iliar with number puzzles. Now they have a chance to become more sophisti-
d by understanding why they work.

The teacher's role in these lessons is very important. Continual encourage-
t and guidance is needed or pupils will just work through the problems
thmetically and omit the translation to algebra. Hints on how to apply
ebra need to be given for each lesson. After the pupils finish a lesson,
ss discussion should include different solutions or a summary of the algebraic
lanation.

In all cases, the teacher will need to be familiar with the algebraic
lanations involved to be sure the pupils have the necessary background.
se are discussed in more detail in the page-by-page commentaries.

These activities are not intended to be used all at one time, but rather at
ious times during the year after the appropriate algebraic skills are obtained.
 algebraic skills emphasized are addition and multiplication of binomials.

Calculators are helpful in some of the lessons. One lesson requires a
endar.

CONNECTORS

```
 0    1   [2]   3    4    5    6    7    8    9

10   11  [12]  13   14   15   16   17   18   19

20   21   22   23   24   25   26   27   28   29

30   31  [32]  33   34  [35]-[36]-37-[38]-[39]

40   41  [42]  43   44   45   46   47   48   49

50   51   52   53   54   55   56  [57]  58   59

60   61   62   63   64   65   66  [67]  68   69

70   71   72   73   74   75   76   77   78   79

80  [81]-[82]-83-[84]-[85]  86  [87]  88   89

90   91   92   93   94   95   96  [97]  98   99
```

1. Four sets of numbers are connected
    For each set add the numbers in the
    squares. Divide the sum by 4.
    What pattern do you see?

Numbers	Sum	Sum ÷ 4

2. Show why the pattern works when the
    connection is a horizontal line.

3. Show why the pattern works when the connection is a vertical line.

4. Experiment with seven connected numbers in a row.

    a. What pattern do you see?

    b. Show why the pattern works.

5. Experiment, as before, with seven connected numbers on a diagonal
    line.

    a. What pattern do you see?

    b. Show why the pattern works.

Connectors

Mathematics teaching objectives:

. Use algebra to make explanations of numerical discoveries.

. Add polynomials.

Problem-solving skills pupils might use:

. Look for patterns.

. Make predictions based upon data.

. Make algebraic explanations.

Materials needed:

. None

Comments and suggestions:

. In order to make the algebraic explanations, pupils need to know how to combine like terms and to divide a monomial by a monomial.

. Much of the activity should be done during class so that you can be available to provide assistance with the algebraic explanations.

Answers:

1.

Numbers	Sum	Sum $\div$ 4
35, 36, 38, 39	148	37
81, 82, 84, 85	332	83
2, 12, 32, 42	88	22
57, 67, 87, 97	308	77

Pattern: The sum divided by 4 is the middle number of the connection.

2. Let N be the middle number. The others are then $N-2$, $N-1$, $N+1$, and $N+2$. Pupils can see that $(N-2) + (N-1) + (N+1) + (N+2) = 4N$ and $4N \div 4 = N$ which is the middle number.

3. Let N be the middle number. The others are then $N-20$, $N-10$, $N+10$, and $N+20$. Pupils can see that $(N-20) + (N-10) + (N+10) + (N+20) = 4N$ and $4N \div 4 = N$ which is the middle number.

4 and 5. a. The sum divided by 6 is the middle number.

b. The explanation is similar to those given in problems 2 and 3.

H's ON THE HUNDREDS GRID

```
 0   1   2  (3)  4  (5)  6   7   8   9

10  11  12 (13)-(14)-(15) 16  17  18  19

20  21  22 (23) 24 (25) 26  27  28  29

30  31  32  33  34  35  36  37  38  39

40  41  42  43  44  45  46  47  48  49

50 (51) 52 (53) 54  55  56  57  58  59

60 (61)-(62)-(63) 64  65  66  67  68  69

70 (71) 72 (73) 74  75  76  77  78  79

80  81  82  83  84  85  86  87  88  89

90  91  92  93  94  95  96  97  98  99
```

The top H is called the 14-H. The bottom H is called the 62-H.

1. Draw the 87-H.   Draw the 55-H.

2. Find the sum of the numbers in the
   a. 14-H ____   b. 62-H ____   c. 87-H ____   d. 55-H ____

3. How can you tell when the H-sum will be even?

4. Find an H with a sum equal to 147.

5. You can use guessing to find the answer to problem 4. But
   there is an easier way. Discover an easy way to find the
   middle number of any H if you know only the sum. Describe
   your method.

<u>H's</u> <u>On</u> <u>The</u> <u>Hundreds</u> <u>Grid</u>

Mathematics teaching objectives:

- . Use algebra to make explanations of numerical discoveries.
- . Add polynomials.

Problem-solving skills pupils <u>might</u> use:

- . Look for patterns.
- . Make predictions based upon data.
- . Make algebraic explanations.

Materials needed:

- . None

Comments and suggestions:

- . The discovery in problem 5 can be made by examining "easier" cases; for example, by studying the results of 11-H, 12-H, 13-H, and 14-H.
- . The algebra needed to explain the discovery is "combining like terms."
- . Much of the lesson should be given during class so that you can be available to provide assistance with the algebraic explanations.

Answers:

1. Pupil drawing.

2. a. 98   b. 434   c. 609   d. 385

3. The middle number in the H is even.

4. 21-H

5. The middle number is $\frac{1}{7}$th of the sum.

6. Let N be the middle number. Then the sum is

$$N + (N + 1) + (N - 1) + (N + 9) + (N - 9) + (N + 11) + (N - 11) = 7N$$

7. 33-H

8. 81-H

9. The sum is 98, the same as before. The algebraic explanation is similar to that given in problem 6.

$$N + (N + 9) + (N - 9) + (N + 10) + (N - 10) + (N + 11) + (N - 11) = 7N$$

6. Show why your discovery in problem 5 works. Begin by letting N be the middle number.

7. Find the largest H with a sum less than 232.

8. Find the smallest H with a sum larger than 565.

9. Turn your paper sideways and draw a sideways 14-H. Find the H-sum of this sideways H. How does it compare to the regular H-sum? Use variables to show why this works.

CALENDAR MATH

Select a month from a calendar.
Select any 4 by 4 grid of 16 days.

1. Find the sum of each diagonal.

   a. _____   b. _____

2. Find the sum of the four
   corners. _____

3. Find the sum of the four
   inside numbers. _____

4. What pattern do you see? _____

5. Use the same 4 by 4 square. Find other combinations of four
   numbers that result in the same magic sum.

6. Show why the pattern occurs. Let N be the first number in
   the 4 by 4 grid.

7. Suppose you know the first number in any 4 by 4 calendar
   square. Without adding, how can you determine the magic sum?

8. For the March 1983 calendar shown at the top, what other
   magic sums are possible?

March 1983

		1	2	3	4	5
6	7	8	9	10	11	12
13	14	15	16	17	18	19
20	21	22	23	24	25	26
27	28	29	30	31		

Calendar Math

Mathematics teaching objectives:

- Use algebra to make explanations of numerical discoveries.
- Add polynomials.

Problem solving skills pupils might use:

- Look for patterns.
- Make predictions based upon data.
- Make algebraic explanations.

Materials needed:

- A calendar

Comments and suggestions:

- You may want to have some old calendars available so that pupils have a chance to work with different months and possibly different years.
- The algebra needed for the explanation is "combining like terms". Thus, the lesson should be used near the time this topic is studied.
- Much of the lesson should be used during class so that you are available to give assistance with the algebraic explanations.

Answers:

1. - 4. The pupils will discover that the sum is the same. (For the calendar shown on the pupil page, the magic sum is 76.)

5. There are many other sets of four numbers that produce the same magic sum. Two of these are given below. Note where these numbers appear on the calendar shown on the pupil page.

(8 + 9) + (29 + 30)   and   (14 + 21) + (17 + 24).

6. Note how the numbers have been represented in the 4 by 4 grid at the right. The magic sum in each case is $4N + 48$.

7. Use the formula, $S = 4N + 48$.

8. 52, 56, 72, and 76

N	N + 1	N + 2	N + 3
N + 7	N + 8	N + 9	N + 10
N + 14	N + 15	N + 16	N + 17
N + 21	N + 22	N + 23	N + 24

THE OUTERS AND THE INNERS

1. Carlos told his teacher about a pattern he had found. Ms. Harris thought others should try it. See if you can find the pattern.

Numbers	Inside Product	Outside Product
3, 4, 5, 6		
5, 6, 7, 8		
9, 10, 11, 12		
15, 16, 17, 18		
50, 51, 52, 53		

   . Use four consecutive numbers like 3,4,5,6.

   . Find the product of the inside numbers. 4 x 5 = 20

   . Find the product of the outside numbers. 3 x 6 = 18

   . Try other numbers. Complete the table.

Pattern: _____

2. Debbie wondered about using four consecutive even numbers like 4, 6, 8, 10. Try five cases.

Numbers	Inside Product	Outside Product

Pattern: _____

3. Show that Carlos' pattern is true. Let N represent the smallest number in the sequence.

4. Show that Debbie's pattern is true.

The Outers And The Inners

Mathematics teaching objectives:

. Use algebra to make explanations of numerical discoveries.

. Multiply binomials.

Problem-solving skills pupils might use:

. Look for patterns.

. Make predictions based upon data.

. Make algebraic explanations.

Materials needed:

. None

Comments and suggestions:

. To explain the pattern, pupils will need to know how to multiply binomials.

. The first three or four problems should be done in class so that you are available to give assistance with the algebraic explanations.

Answers:

1.

	Inside Product	Outside Product
3, 4, 5, 6	20	18
5, 6, 7, 8	42	40
9, 10, 11, 12	110	108
15, 16, 17, 18	272	270
50, 51, 52, 53	2652	2650

Pattern: The inside product is 2 greater than the outside product.

2. For four even consecutive numbers, the inside product is 8 greater than the outside product.

3. Let N be the smallest number. Then the inside product is

$$(N + 1)(N + 2) = N^2 + 3N + 2$$

and the outside product is

$$N(N + 3) = N^2 + 3N .$$

By subtracting the two products we see that the difference is 2.

$$(N^2 + 3N + 2) - (N^2 + 3N) = \underline{2}$$

4. The algebraic proof is similar to that given for problem 3. This time the numbers can be represented as N, N + 2, N + 4, and N + 6. Some pupils may feel that the smallest number should be represented by 2N since the numbers are even. Such representation can be used, but it is not necessary for problems like these.

5. Do the same experiment but this time use four consecutive odd numbers.

   a. What pattern do you see?

   b. Show why the pattern works.

6. Experiment, as before, with four consecutive numbers that differ by 3. (For example, 5, 8, 11, and 14)

   a. What pattern do you see?

   b. Show why the pattern works.

7. Experiment with four consecutive numbers that differ by 4.

   a. What pattern do you see?

   b. Show why the pattern works.

8. Suppose you were to use four consecutive numbers that differ by 5.

   a. What do you predict the result to be? (Make the prediction without experimenting.)

   b. Now, set up an experiment to check your prediction.

   c. Show why the pattern works.

<u>The</u> <u>Outers</u> <u>and</u> <u>Inners</u>

Answers:

5. The results are the same for odd numbers as they were for evens.

6. a. The difference is 18.

    b. The algebraic proof is similar to that given for problem 3.

7. a. The difference is 32.

    b. The algebra is similar to problem 3.

8. a. 50   This prediction can be made by observing a pattern in the
            differences: 2, 8, 18, 32, <u>50</u>   or   2 · 1, 2 · 4, 2 · 9, 2 · 16, <u>2 · 2</u>

    b. Pupil experiment.

    c. The algebra is similar to problem 3.

IN A ROW

0	1	2	3	4	5	6	7	8	9
10	11	12	13	(14	15	16)	17	18	19
20	21	22	23	24	25	26	27	28	29
30	(31	32	33)	34	35	36	37	38	39
40	41	42	43	44	45	46	47	48	49
50	51	52	53	54	55	56	57	58	59
60	61	(62	63	64)	65	66	67	68	69
70	71	72	73	74	75	76	77	78	79
80	81	82	83	84	85	86	(87	88	89)
90	91	92	93	94	95	96	97	98	99

1. Look at the circled numbers. For each, multiply the first and the last. Square the middle. Record in the table.

Numbers	First Times Last	Middle Squared
14, 15, 16		
31, 32, 33		
62, 63, 64		
87, 88, 89		

a. What pattern do you see? _____

b. Show why the pattern works. Let N be the middle number.

2. Choose five consecutive numbers. Multiply the first and the last. Square the middle. Record. Repeat two more times.

Numbers	First Times Last	Middle Squared

a. What pattern do you see? _____

b. Show why the pattern works.

-99-

© PSM 82

Mathematics teaching objectives:

- Use algebra to make explanations of numerical discoveries.
- Multiply binomials.

Problem-solving skills pupils <u>might</u> use:

- Look for patterns.
- Make predictions based upon data.
- Make algebraic explanations.

Materials needed:

- None

Comments and suggestions:

- To explain the pattern, pupils need to know how to multiply binomials.
- Much of the activity should be done during class so that you can provide assistance with the algebraic explanations.
- After the lesson is completed you may wish to discuss shortcuts for multiplying numbers like 23 X 17. Note that this can be thought of as $(20 + 3)(20 - 3)$ or $20^2 - 3^2$. Similarly, 55 X 45 can be thought of as $(50 + 5)(50 - 5)$ or $50^2 - 5^2$. The algebraic basis for this shortcut is $(a + b)(a - b) = a^2 - b^2$.

Answers:

1.

Numbers	First times Last	Middle Squared
14, 15, 16	224	225
31, 32, 33	1023	1024
62, 63, 64	3968	3969
87, 88, 89	7743	7744

    a. The product of the first and last number is one less than the square of the middle number.

    b. Let N be the middle number. N - 1 is the first number. N + 1 is the second number.
$$N^2 - (N - 1)(N + 1) = 1$$

2. a. The product of the first and last number is four less than the square of the middle number.

    b. The algebraic proof is similar to 1 b.

3. a. The product of the first and last number is 9 less than the square of the middle number.

    b. The algebraic proof is similar to 1 b.

4. 49 (or 7^2)

3. Choose seven consecutive numbers. Multiply the first and the last. Square the middle. Record. Repeat two more times.

Numbers	First Times Last	Middle Squared

   a. What pattern do you see? _____

   b. Show why the pattern works.

4. Suppose you chose 15 consecutive numbers. What would be the difference between the middle number squared and the product of the first and last?

STRANGE HAPPENINGS

```
 0    1    2    3    4    5    6    7    8    9

10   (11)  12  (13)  14   15   16   17   18   19

20   21   22   23   24   25  (26)  27  (28)  29

30  (31)  32  (33)  34   35   36   37   38   39

40   41   42   43   44   45  (46)  47  (48)  49

50   51   52   53   54   55   56   57   58   59

60   61   62   63  (64)  65  (66)  67   68   69

70   71   72   73   74   75   76   77   78   79

80   81   82   83  (84)  85  (86)  87   88   89

90   91   92   93   94   95   96   97   98   99
```

For each of the above 3 by 3 arrays,

. record the products of the
  two pairs of numbers
  connected by a line.

. subtract the smaller product
  from the larger product. Record.

What pattern do you see? _____

Product	Product	Difference

SM 82

<u>Strange</u> <u>Happenings</u>

Mathematics teaching objectives:

  . Use algebra to make explanations of numerical discoveries.

  . Multiply binomials.

Problem-solving skills pupils <u>might</u> use:

  . Look for patterns.

  . Make predictions based upon data.

  . Make algebraic explanations.

Materials needed:

  . None

Comments and suggestions:

  . To explain the pattern, pupils will need to know how to multiply binomials

  . Much of the activity should be done during class so that you are
    available to give assistance with the algebraic explanations.

  . Some possible extensions: explore a 3 by 2, 3 by 4, 3 by 5, etc.;
    explore a 4 by 4, 4 by 5, 4 by 6, etc.

Answers:

1.

Product	Product	Difference
363	403	40
1248	1288	40
5504	5544	40

Pattern: The difference
between the products
is 40.

2. a. n + 2, n + 20, n + 22

   b. $(n + 2)(n + 20) - (n)(n + 22)$
      $n^2 + 22n + 40 - (n^2 + 22n) = 40$

3. a.

Array Size

	2 by 2	3 by 3	4 by 4	5 by 5
Difference	10	40	90	160

  b. The algebraic proofs are similar to that given in problem 2.

Strange Happenings (cont.)

2. Let <u>n</u> be the smallest number in a 3 by 3 array.

    a. How would the other three numbers be represented?

    b. Use these representations to show that the pattern
        in problem 1 works.

3. Experiment with arrays of different sizes.

    a. Record the difference between the products.

    b. Show why the patterns work.

	Array Size			
	2 by 2	3 by 3	4 by 4	5 by 5
Difference				

4. Suppose you were to use a 10 by 10 array.

    a. What do you predict the result (the difference) to be?
        Make the prediction without experimenting.

    b. Show why this pattern would work.

Answers:

4. a. The difference in a 10 by 10 array is 810. This prediction can be made by observing a pattern in the differences:

$40 - 10 = \underline{30}$, $90 - 40 = \underline{50}$, $160 - 90 = \underline{70}$, . . .

Another pattern that pupils might observe:

$10 = 1^2 \times 10$ (2 by 2)

$40 = 2^2 \times 10$ (3 by 3)

$90 = 3^2 \times 10$ (4 by 4)

$160 = 4^2 \times 10$ (5 by 5)

$\vdots$

$810 = 9^2 \times 10$ (10 by 10)

b. The algebraic explanation is similar to that given in problem 2.

Algebra

IV. EQUATION SOLVING

Most pupils have been introduced to equation solving in courses prior to
algebra. Usually the equations were simple and were solved by using formal
equation-solving procedures. In the first three lessons, more difficult equations are used and informal procedures are encouraged. Explicit use is made
of the problem-solving skills, guessing and checking, making charts, and
looking for patterns. The remaining pages in this section are teacher suggestions for methods that might be used to bridge the gap between informal and
formal methods for solving first-degree equations with one variable. The
development and practice of formal equation-solving skills will need to be
provided from a textbook or worksheets.

In an earlier section, variables were used in expressions (polynomials) such
as 2a + b and n^2 + 3n + 2. In these cases the variables a, b, and n are symbols
from some set of numbers. Usually, up to this point in algebra, the numbers are
integers and occasionally rational numbers. The value for the expressions are
determined by using these numbers without restrictions.

The informal procedures suggested on the pupil pages afford opportunities to
examine the use of variables in equations. In the first lesson, a "window" plays
the role of the variable and a "suitcase full of numbers" is that some set of
numbers. The problems suggest that a guess-and-check strategy be used to find
numbers which make the equations true statements. Each selection encourages
pupils to check by going back to the equation and reviewing its meaning. The
solution (often called truth value for the variable) is restricted to one answer.

In the second lesson, a window is used on both sides of the equal sign. Each
window serves as a symbol for the same variable. Two equations in this lesson do
not have a single solution--one equation has no solution, the other infinitely
many.

In the third lesson, the window is replaced by n . The equations have
either one, infinitely many, or no solution. A goal of the lesson is to convince
pupils that guess-and-check strategies are inefficient and there must be better
ways to solve equations. Also, the rational expressions $\frac{n}{n + 1}$ and $\frac{10}{x}$ are used
in some of the equations calling for solutions. A restriction needs to be made.
Zero must be taken from the "suitcase full of numbers" if any number in the
"suitcase" is to be used in evaluating these expressions. This type of restriction should be emphasized whenever the opportunity arises any time during the
year.

ONE WINDOW

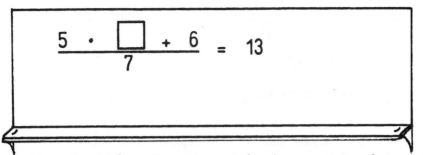

$$\frac{5 \cdot \boxed{}}{7} + 6 = 13$$

The pupils in Mr. Smith's class need to have a good imagination. When they came to class they found the above equation on the chalkboard. Mr. Smith picked up an imaginary saw and pretended to cut out the $\boxed{}$ shape.

"Don't forget to clean up the sawdust," said Jeff. "And don't forget to tell the custodian that we need a new chalkboard."

Mr. Smith said that Miss Adams, who teaches in the room next door, has a suitcase full of numbers. She's agreed to hold up numbers so they can be seen through the hole. We need to decide whether the number is too small, too large, or just right.

Class Exercises

1. Miss Adams first used a "10." Is this number too large or too small?

2. The next number used was a "20." Is this too large or too small? Is the correct solution closer to "20" or "10?"

3. Can you suggest a number for Miss Adams to use that will be "just right?" What is the solution to the equation?

Exercises

Below are other equations that Mr. Smith wrote on the board. For each of them, decide what number is the correct solution.

1. $17 \cdot \boxed{} + 7 = 211$

2. $8 \cdot \boxed{} - 29 = 299$

3. $12 \cdot \boxed{} + 37 = 1$

4. $\dfrac{3 \cdot \boxed{}}{4} + 5 = 29$

One <u>Window</u>

Mathematics teaching objectives:

- Solve equations using a guess-and-check strategy.
- Compute mentally.
- Create equations with a given solution.

Problem-solving skills pupils <u>might</u> use:

- Guess and check.
- Record solution possibilities or attempts.
- Look for patterns.
- Determine limits.

Materials needed:

- None

Comments and suggestions:

- Much of the lesson should be done in class so the teacher can provide assistance as needed.
- A guess, check, and refine strategy should be emphasized. No formal equation-solving methods should be discussed at this time. Nevertheless, pupils should be encouraged to search for methods or processes which make equation solving more efficient.
- The "hole in the wall" approach to variables encourages the pupil to examine the relationships stated in the equation. This leads to a better understanding of its structure.
- Encourage pupils to record their guesses. This is important to the next step of organizing and analyzing the guesses.
- Pupils should have sufficient time to gain some proficiency in using their guess, check, and refine strategies.

Answers:

Class Exercises

1. Too small
2. Too large - closer to 20
3. Suggestions will vary. Solution is 17.

Exercises

1. 12	2. 41	3. -3	4. 32
5. 4	6. 120	7. 8	8. -4
9. 10	10. 11 or -11	11. -5	12. 0

13. Responses will vary. One possibility:

$$\frac{3 \cdot \Box + 2}{4} = 5$$

One Window (cont.)

5. $9(\boxed{} - 6) = {}^{-}18$

9. $\dfrac{2 \cdot \boxed{} + 1}{3} - 3 = 4$

6. $\dfrac{\boxed{}}{10} + 2 = 14$

10. $\boxed{}^2 - 12 = 109$

7. $23 = 4 \cdot \boxed{} - 9$

11. $3 = \dfrac{6}{\boxed{} + 7}$

8. $\dfrac{48}{\boxed{}} + 14 = 2$

12. $\dfrac{{}^{-}4 \cdot \boxed{} - 6}{3} = {}^{-}2$

13. The solution to each of these equations is 6.

$5 \cdot \boxed{} - 1 = 29$

$\dfrac{3 \cdot \boxed{} + 12}{10} = 3$

Write 2 more equations that also have a solution of 6.

TWO WINDOWS

$$5 \cdot \boxed{} - 1 = 2 \cdot \boxed{} - 7$$

It was necessary for Mr. Smith to cut out two "windows" in this equation. And this time Miss Adams needed a duplicate set of numbers. The agreement was that whatever appears in one window must also appear in the other.

Class Exercises

1. Miss Adams first used an "8" in each window. Is this number too large or too small, or can you tell?

2. Next, Miss Adams used a "10" in each window. Is the correct solution closer to "10" or "8," or can you tell?

3. The next number used was a "12." Is this guess any better than before, or can you tell?

4. The class suggested more and more numbers for Miss Adams to try. But none of them seemed to be very helpful. Mr. Smith suggested using a chart. Sometimes a chart is helpful in revealing certain patterns.

Guess	Left Side	Right Side
8	39	9
10	49	13
12	59	17

   a. Study the chart. What patterns do you see?

   b. Are the guesses getting closer to the solution or further away?

   c. Add more numbers to the chart. What is the solution to the equation?

5. Use a chart to help you solve $2 \cdot \boxed{} + 32 = 6 \cdot \boxed{} - 28.$

Two Windows

Mathematics teaching objectives:

- Solve equations using informal methods.
- Compute with integers.
- Organize data into a chart form.

Problem-solving skills pupils _might_ use:

- Guess and check.
- Make a systematic list.
- Look for a pattern.
- Create new problems.

Materials needed:

- None

Comments and suggestions:

- Do this lesson in class. Provide assistance as needed.
- Help from persons outside the class could work against the purpose of this lesson.
- By using two "windows," the equation presents a more complex situation to the pupil. The guesses now require greater organization. The use of a chart helps the pupil to organize and refine the guesses. The analysis of the results in the chart provides a deeper understanding of the relationships in the equation. Encourage the use of the chart at this stage of development even though pupils may want to "shortcut" it. It might be helpful to provide a ditto of blank chart forms for the pupils.
- Discuss the patterns that appear in the charts. Searching for and discovering patterns will also strengthen understandings of the structure of equations.
- It is important for pupils to see that some equations will have no solution, while others have infinitely many solutions. Often pupils seem to progress through mathematics under the impression that every problem has a single solution.

Answers:

See page 118.

Two Windows (cont.)

Exercises

Solve each equation. Remember, you must use the same number in each window.

1. $5 \cdot \boxed{} = 2 \cdot \boxed{} + 57$

2. $7 \cdot \boxed{} + 15 = 6 \cdot \boxed{} + 14$

3. $5 \cdot \boxed{} + 2 = 17 \cdot \boxed{} + 2$

4. $3 \cdot \boxed{} + 7 = 4 + 3 \cdot \boxed{}$

5. $\dfrac{\boxed{} + 5}{2} + \boxed{} = 19$

6. $4\,(\boxed{} + 5) = 4 \cdot \boxed{} + 20$

7.   Try to solve this "monster" that Mr. Smith created.

$$\dfrac{2 \cdot \boxed{} - 3 \cdot \boxed{}}{5} - 6 = 3\,(\boxed{} - 2)$$

8.   Create a "monster" of your own. Use at least two windows. Be sure you know the solution before giving it to someone else to solve.

Answers:

Class Exercises

1. Can't tell.

2. Probably still can't tell.

3. Probably still can't tell (unless someone sees a pattern).

4. a. Responses will vary.

    Some possible patterns:

    Left side increases by 10 each time.

    Right side increases by 4 each time.

    Differences increase by 6 each time.

    b. Further away.

    c. $^-2$

5. Possible chart:

Guess	Left Side	Right Side
10	52	32
8	48	20
12	56	44
14	60	56
15	62	62

Exercises

1. 19

2. $^-1$

3. 0

4. No solution possible.

5. 11

6. The solution can be any number. This is an example of the distributive property.

7. 0

8. Encourage each pupil to develop a "monster" equation. It provides a valuable experience in equation structuring. Responses will vary.

    One possibility: $3 \cdot \square + 4 = \dfrac{2 \cdot \square}{3} + 25$

A TRICKY PROBLEM

Felix is the class trickster. He created this "monster" and gave
it to the class to solve.

$$\frac{11N + 3}{2} + 1 = \frac{3N + 1}{2} + 15$$

Many students thought
he was tricking them
by using N's rather than windows. "That's not what I had in mind
at all," said Felix. "You do the same as before. Each N must be
replaced by the same number." So the class proceeded to solve the
equation. They were determined to find out why Felix thought his
problem was so "tricky."

Class Exercises

1. Why did Felix think his problem was a tricky one?

2. What is the closest whole number solution to Felix's equation?

3. Find the closest whole number solution to $\frac{10A + 3}{6} - 7 = 17$.

Exercises

Solve the following equations. If a solution is a fraction, find
the closest integer.

1. $\frac{8N + 3}{4} - 6 = 1$

2. $\frac{N + 5}{2} + N = 19$

3. $7 + Y = 7Y + 36$

4. $\frac{3 + N}{5} = N + 2$

5. $3(N + 4) + 3(N - 4) = 50$

6. $^-3 + Y = ^-Y - 3$

7. $\frac{N}{N + 1} = 1$

8. $\frac{X}{10} + \frac{10}{X} = 10$

A <u>Tricky</u> <u>Problem</u>

Mathematics teaching objectives:

    . Solve equations using informal methods.

    . Compute with integers.

    . Round numbers.

Problem-solving skills pupils <u>might</u> use:

    . Guess and check.

    . Make a systematic list.

    . Look for a pattern.

Materials needed:

    . None

Comments and suggestions:

    . In this lesson, the equations have become more difficult to solve by a guess and check procedure. Hopefully, by the end of the lesson, pupils will recognize a need for learning equation-solving procedures other than <u>guess</u> <u>and</u> <u>check</u>. Perhaps some pupils have even discovered some of their own methods.

    . Remind pupils that organizing their guesses in a chart makes guessing more efficient.

    . Encourage pupils to search for the <u>closest</u> solution. Remind them to "trap" the solution between two integers and then decide which integer it's closest to.

    . Have pupils give reasons why the variables in the expressions $\frac{N}{N+1}$ and $\frac{10}{X}$ must be restricted (see Overview comments page).

Answers:

<u>Class</u> <u>Exercises</u>

1. Because the solution is fractional and difficult to find by guess and check.

2. 3

3. 14

<u>Exercises</u>

1. 3 (rounded)

2. 11

3. ⁻5 (rounded)

4. ⁻2 (rounded)

5. 8 (rounded)

6. 0

7. No solution

8. Both 100 and 1 are good approximations. Each gives a result of $10\frac{1}{10}$. An even better approximation is 99.

SOME PROCEDURES FOR SOLVING EQUATIONS
(Ideas For Teachers)

Usually at this time pupils are interested in learning a more efficient equation-solving method than guess and check. If pupils have better methods, let them explain their procedures.

Several equation-solving procedures are outlined on the following pages. You may wish to discuss one or more of them with the class. The procedure most often discovered by pupils is the "cover-up" method (see page 123). Other methods you may wish to demonstrate are "Machine Hook-ups" (page 125) and "Unwrapping The Package" (page 127).

These three methods are suitable for first-degree equations with the variable on only one side of the equation. Eventually, however, some general equation-solving properties need to be developed in order to solve all first-degree equations. These properties usually are referred to as the addition, subtraction, multiplication and division properties of equality. There are various ways these properties can be presented. One way is to illustrate them by using "Money Bag" activities (page 129).

COVER-UP

Use a piece of paper to cover parts of an equation while pupils reason out what value the covered-up part must have. The equation below and appropriate questions illustrate the method.

Equation	Pupils See	Question
A. $\dfrac{3N - 2}{5} + 7 = 15$		
B. $\dfrac{3N - 2}{5} + 7 = 15$	$\boxed{8} + 7 = 15$	What must the covered-up part be to make the equation true? [Answer: 8]
C. $\dfrac{3N - 2}{5} = 8$	$\dfrac{\boxed{40}}{5} = 8$	What must the covered-up part be to make the equation true? [Answer: 40]
D. $3N - 2 = 40$	$\boxed{42} - 2 = 40$	What must the covered-up part be to make the equation true? [Answer: 42]
E. $3N = 42$	$3 \times \boxed{14} = 42$	What must the covered-up part be to make the equation true? [Answer: 14]
F. $N = 14$		

You might want to actually write the value on the piece of paper used to do the cover-up. The dotted numerals show this technique.

PSM 82

Pupils first need to work problems involving a variety of hook-ups like those given below.

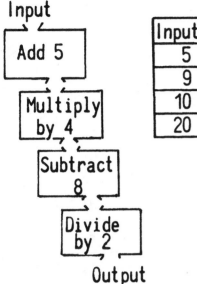

Input	Output
5	?
9	?
10	?
20	?

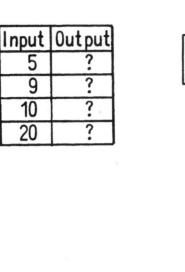

Input	Output
15	?
8	?
9	?
20	?

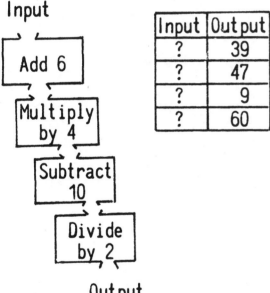

Input	Output
?	39
?	47
?	9
?	60

Next, have pupils work problems which require them to find the input when the output is given. Pupils should discover that working backwards is an efficient method. (For example, first multiply by 2, then add 10, then divide by 4, and finally subtract 6.)

Have pupils solve the equation $\frac{(n + 6) \cdot 4 - 10}{2} = 60$. Show how the hook-up in B is directly related to this equation.

Have pupils devise their own machine hook-ups to solve equations like $\frac{3n - 2}{5} + 7 = 15$.

E. Have pupils attempt to devise a machine hook-up for solving an equation like $3n + 2 = 2n + 3$.

Of course, this equation is different from the others and this kind of activity can be used to motivate the need for another equation-solving method.

UNWRAPPING THE PACKAGE

Pupils first need to visualize the steps in wrapping a package. Have them determine the order of these steps.

. Tie the ribbon around the box.
. Place the gift in the box.
. Wrap the paper around the box.
. Put the lid on the box.

To unwrap the package the steps must be (a) <u>undone</u> and (b) <u>in the reverse order</u> as wrapping the package.

For example, to "wrap" the simple equation $y = 10$, we might do these steps:

a. $\quad\quad\quad y = 10$

b. $\quad\quad\quad 4y = 40$

c. $\quad\quad 4y - 8 = 32$

d. $\quad\quad \dfrac{4y - 8}{2} = 16$

e. $\quad \dfrac{4y - 8}{2} + 5 = 21$

To "unwrap" $\dfrac{3n - 2}{5} + 7 = 15$ the steps must be (a) <u>undone</u> and (b) <u>in the reverse order</u>.

a. $\quad\quad\quad\quad\quad \dfrac{3n - 2}{5} + 7 = 15$

b. Subtract 7 $\quad\quad \dfrac{3n - 2}{5} = 8$

c. Multiply by 5 $\quad 3n - 2 = 40$

d. Add 2 $\quad\quad\quad\quad 3n = 42$

e. Divide by 3 $\quad\quad\quad n = 14$

So the solution to $\dfrac{3n - 2}{5} + 7 = 15$ is $n = 14$.

MONEY BAGS

Each money bag below contains the same number of pennies.
The total amount on the left side is the same as the
total amount on the right side.

EACH BAG CONTAINS _____ ¢

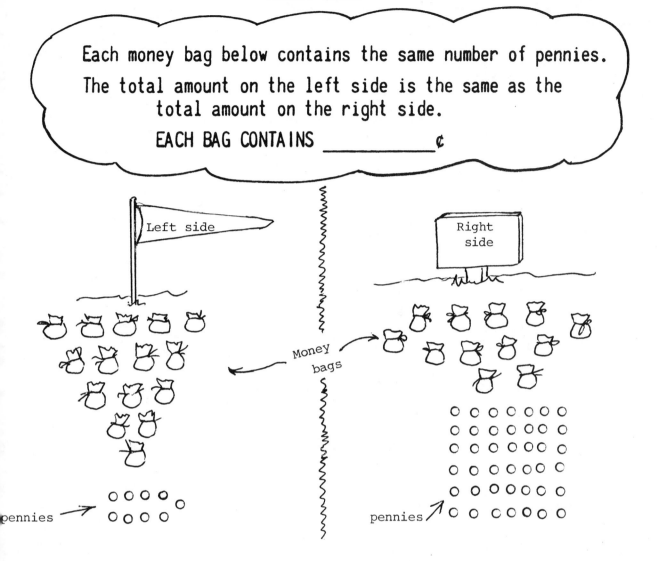

Comments and suggestions:

- Pupils should be discouraged from using guess-and-check procedure on this problem.

- The obvious statement   --   "The amount in each bag remains the same if the same
                                number of bags or pennies is taken from each side."
  is analogous to   --   "If the same value is subtracted from each side of an
                          equation, the new equation has the same solution as the
                          original equation.

- Here is one sequence of steps   1st - Take 12 bags from each side.
  that can be used: 2nd - Take 9 pennies from each side.
                                    3rd - Divide each set by 3.

- Algebraically, these $15B + 9 = 12B + 42$
  steps take this form: Subtr. 12B   Subtr. 12B
                                      $3B + 9 = 42$
                           Subtr. 9 $3B = 33$ Subtr. 9
                           Divide by 3 $B = 11$ Divide by 3

- Problems like this are easy to create. Perhaps pupils could create some of
  their own and have classmates check them.

Algebra

V. WORD PROBLEMS

V. WORD PROBLEMS

Word problems frustrate almost everyone. Teachers find them difficult to teach and pupils dislike them. An approach is used here which may help alleviate some of this frustration. The initial lesson encourages pupils to use a guess-and-check strategy. The guesses lead to the use of charts and finally to the more traditional format in the fourth lesson. In a sense, this section ends where most textbook treatment of word problems begins. Each lesson is dependent upon the previous one and should be done in the order given. However, you may choose to provide similar problems if more practice is needed with the strategies emphasized in each lesson.

Since the focus of the lessons here is on the translation process, pupils should have some familiarity with variables and equation solving before starting this section. It should be used prior to the textbook lessons on word problems.

PUZZLE PROBLEMS

Many puzzles, like the one at the right, can be solved by guessing. Have you ever guessed the answer to a problem on the first try? Sometimes this happens. Usually, however, you need to refine your guess and try again.

Jenny has some dimes and nickels in her purse.
She has four more dimes than nickels.
The total value is $4.00.
How many of each coin does Jenny have?

Suppose your first guess to Jenny's puzzle is ten nickels. Is this guess too large or too small? What would be your next guess? Try other guesses until you find the answer to Jenny's puzzle.

Use guessing to help you try to solve these problems. Some of them may not be possible to solve. If so, write N.P.

1. One number is 28 larger than another. Their sum is 420. What are the two numbers?

2. One number is five times as large as another. Their sum is 900. What are the two numbers?

3. These two sets are examples of consecutive numbers:

   5, 6, 7, 8, 9, 10   51, 52, 53, 54, 55

   Find four consecutive numbers whose sum is 102.

4. Find five consecutive numbers whose sum is 175.

5. These are examples of consecutive odd numbers:

   9, 11, 13, 15, 17   39, 41, 43, 45, 47, 49

   Find four consecutive odd numbers whose sum is 240.

Puzzle Problems

Mathematics teaching objectives:

. Guess and check an answer to a word problem.

. Extract pertinent data and relationships from a word problem.

Problem-solving skills pupils <u>might</u> use:

. Guess and check.

. Record solution attempts.

. Look for patterns.

Materials needed:

. None

Comments and suggestions:

. Suggest the use of a guess-and-check strategy for solving these puzzle problems, but do not discourage the use of other methods. Guessing and checking requires pupils to read the puzzle several times, leading them to a better understanding of the problem.

. In the word problems in this section, emphasize that the thinking process they use to get the answers to puzzles is more important than the answers themselves.

. Have pupils work on the problems independently. Be as nondirective in your assistance as you can manage.

. Stress guess and check during class discussions deferring the discussion of the more systematic procedures you observed to subsequent lessons.

. The lesson should be done in class to avoid the premature instruction from others on formal algebraic procedures.

Answers:

1. 224 and 196   2. 750 and 150   3. 24, 25, 26, 27

4. 33, 34, 35, 36, 37   5. 57, 59, 61, 63   6. 30 cm, 30 cm, and 40 cm

7. 26 cm and 52 cm   8. 62°, 62°, 56° 9. 51°, 58°, and 71°

10. Mary is 24, Jean is 12, Hazel is 17 years old.

11. Not possible.

12. 35 canaries and 25 cats. (Remind the pupils that all the "critters" are normal.)

13. Not possible.

Puzzle Problems (cont.)

6. The perimeter of a figure is the "distance around."
   Suppose that -
   . two sides of a triangle are equal.
   . the third side is 10 cm greater than each of the other two.
   . the perimeter is 100 cm.
   Find the length of each side.

7. The perimeter of a rectangle is 156 cm. The length is twice
   as long as the width. Find the length and width.

8. The sum of the three angles of a triangle is always 180^0.
   Suppose that -
   . two angles of a triangle are equal.
   . the third angle is 6^0 less than each of the other two.
   How large is each angle?

9. One angle of a triangle is 7^0 more than the smallest angle.
   The third angle is 20^0 more than the smallest. How large is
   each angle?

10. Mary is twice as old as Jean and 7 years older than Hazel.
    The sum of their ages is 53. How old is each?

11. Jim has $6 in dimes and nickels. He has 5 more nickels than
    dimes. How many dimes does he have?

12. Mrs. Howard raises cats and canaries. Altogether the
    animals have 170 feet and 60 heads. How many cats and
    how many canaries does she have?

13. Mr. Farmer raises ducks and cows. The animals have a
    total of 20 heads and 55 feet. How many ducks and
    how many cows does he have?

USING CHARTS

Wayne used a chart to help him solve a number puzzle. The chart below shows how he organized his guesses.

1ST NUMBER	2ND NUMBER	3RD NUMBER	TOTAL
10	20	23	53 (too small)
20	40	43	103 (too large)
15	30	33	78 (too large)
14	28	31	73 (just right)

1. Study the chart carefully. Write down what you think was the puzzle that Wayne solved.

2. Study the chart below. Write down what you think the problem might have been.

SID'S AGE	KIM'S AGE	TANA'S AGE	TOTAL
12	10	30	52 (too large)
8	6	18	32 (too large)
5	3	9	17 (too small)
6	4	12	22 (just right)

Using Charts

Mathematics teaching objectives:

- Extract pertinent data and relationships from a word problem.
- Organize data from a word problem into a chart.

Problem-solving skills pupils might use:

- Guess and check.
- Make and use a table.
- Look for a pattern.
- Invent a problem which can be solved by certain procedures.

Materials needed:

- None

Comments and suggestions:

- Guess-and-check strategies are still used in this lesson. However, making and using tables is the primary focus since this will be used as a basis for introducing equation-solving procedures in the next lesson.
- Suggest another column be added to the chart to record "how far off" the guesses were.
- Allow enough independent study time for the lesson to be completed in class. As in the last lesson, premature instruction from others on formal algebraic procedures works against the purpose of this lesson.

Answers:

1. Responses will vary. One possibility: The sum of three numbers is 73.
   The second number is double the first.
   The third number is three more than the second.
   What are the numbers?

2. Responses will vary. One possibility: Kim is two years younger than Sid.
   Tana is three times as old as Kim.
   The sum of their ages is 22 years.
   How old is each?

Using Charts (cont.)

3. Read the puzzle at the right. Use the first row of the chart below to show that a guess of 5 quarters is too small. Try other guesses. Record them in the chart. What is the answer to Eric's puzzle?

Eric has some quarters and nickels in his pocket.
He has 8 more nickels than quarters.
The total value is $3.70.
How many of each coin does Eric have?

Quarters	Cents in Quarters	Nickels	Cents in Nickels	Total amount in Cents
5	125			

4. Use a chart to help you solve these problems.

a. A 99-inch rope is cut into two parts. One part is 29 inches longer than the other. How long is each piece?

One Part	Other Part	Total

b. The perimeter of a triangle is 108 cm. One side is twice as long as the shortest side. The third side is 3 times as long as the shortest. How long is each side?

c. In a school election, Cindy received 135 more votes than Alice, and Alice received twice as many votes as Juanita. If 490 votes were cast for the three candidates, how many did Juanita receive? (First, try a guess of 100.)

Answers:

3.

Quarters	Cents in Quarters	Nickels	Cents in Nickels	Total amount in Cents	
5	125	13	65	190	(too small)
⋮	Guesses will vary.			⋮	
11	275	19	95	370	(just right)

4. a.

One Part	Other Part	Total Length
⋮	Guesses will vary.	⋮
35 inches	64 inches	99 inches

b.

Shortest Side	Longest Side	Other Side	Perimeter
⋮	Guesses will vary.		⋮
18 cm	54 cm	36 cm	108 cm

c.

Juanita	Alice	Cindy	Total	
100	200	335	635	(too large)
⋮	Guesses will vary.		⋮	
71	142	277	490	

GUESSES - CHARTS - EQUATIONS

> The sum of three numbers is 99.
>
> The second number is 5 times as large as the first.
>
> The third number is 4 more than the second.
>
> Find the smallest number.

1. Read the problem carefully. Try some guesses. Put them in the chart below.

   Why is this problem difficult to solve by guessing?

1st number	2nd number	3rd number	Sum

2. Between what two numbers is the answer to the puzzle?

3. You could continue to make guesses. But since the answer is a fraction, a guessing procedure could take a long time. Study the chart below. Notice that a box is used to keep track of a guess of 10.

1st number	2nd number	3rd number	What the sum should be
$\boxed{10}$	$5 \cdot \boxed{10}$	$5 \cdot \boxed{10} + 4$	99

Add a guess of 9 to the chart. Put it in a box and use the **same** form as was used for the first guess of 10.

Guesses - Charts - Equations

Mathematics teaching objectives:

- Extract pertinent data and relationships from word problems.
- Derive equations from organized charts.
- Solve equations.

Problem-solving skills pupils might use:

- Guess and check.
- Make a systematic list.
- Use mathematical symbols to describe situations.

Materials needed:

- None

Comments and suggestions:

- This is the most important lesson in this section. The transition is made from the chart to the equation and needs to be developed carefully in class. Help pupils with the "boxing" process.
- The temptation for pupils to use shortcuts is strong at this time. However, the guessing strategy and the use of a chart needs reinforcement if they are to use these skills in the word problems they will encounter later.
- The first problem requires fractions as answers and, therefore, motivates the need for another strategy--building equations for word problems.

Answers:

1 and 2.   Guesses will produce no whole number solutions. The number is between 8 and 9.

3.

1st number	2nd number	3rd number	What the sum should be
$\boxed{10}$	5 · $\boxed{10}$	5 · $\boxed{10}$ + 4	99
$\boxed{9}$	5 · $\boxed{9}$	5 · $\boxed{9}$ + 4	99

4. The information in the first two rows of the chart can be written as

$$\boxed{10} + 5 \cdot \boxed{10} + 5 \cdot \boxed{10} + 4 \overset{?}{=} 99$$

$$\boxed{9} + 5 \cdot \boxed{9} + 5 \cdot \boxed{9} + 4 \overset{?}{=} 99$$

Neither of these guesses give the correct answer. But the pattern of the problem is established. Instead of continuing with more guesses, an equation can be written.

$$N + 5N + 5N + 4 = 99$$

Solve this equation for N. What is the answer to the puzzle given at the first of the lesson?

5. Sally and Bev are long distance runners. They leave at the same time and same point and run in opposite directions. Suppose Sally averages 8 miles per hour and Bev averages 10 miles per hour. How soon will they be 25 miles apart?

Study the chart below. Notice that boxes are used to keep track of the first guess.

Number of Hours	Sally's Distance	Bev's Distance	What the Total Distance Should Be
$\boxed{3}$	$\boxed{3} \cdot 8$	$\boxed{3} \cdot 10$	25

Answers:

4. $N = 8 \frac{7}{11}$

5. a. Too large

b.

Number of Hours	Sally's Distance	Bev's Distance	What the Total Distance Should Be
3	3 · 8	3 · 10	25
2	2 · 8	2 · 10	25

Too large.

c. $N \cdot 8 + N \cdot 10 = 25$

d. $N = 1 \frac{7}{18}$ hours

6. a.

Guesses will vary.

First Angle	Second Angle	Third Angle	Sum of Angles
☐	☐ + 8	3 · ☐	180°

$$N + N + 8 + 3N = 180°$$

$$N = 34 \frac{2}{5}°$$

b.

Guesses will vary.

Smaller Number	Twice Smaller	Larger Number	Sum
☐	2 · ☐	2 · ☐ + 3	77

$$N + 2N + 3 = 77$$

$$N = 24 \frac{2}{3}$$

    a. Is a guess of 3 hours too large or too small?

    b. Add a guess of 2 hours to the chart. Use the same form as before. Is 2 hours too large or too small?

    c. Write an equation that can be used to solve the problem. (Use only the last three columns of the chart.)

    d. Solve the equation. What is the answer to the problem?

6. In each problem –

    . make two guesses. Organize your guesses in a chart.

    . use your chart to help you write an equation.

    . solve the problem.

    a. The second angle of a triangle is 8^0 more than the first. The third angle is 3 times as large as the first. What is the size of the smallest angle?

Answers:

6. c.

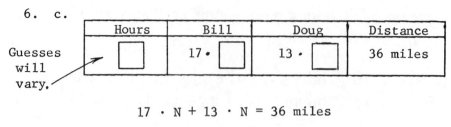

Guesses will vary.

Hours	Bill	Doug	Distance
☐	17 · ☐	13 · ☐	36 miles

$$17 \cdot N + 13 \cdot N = 36 \text{ miles}$$

$$N = 1\frac{1}{5} \text{ hours}$$

d.

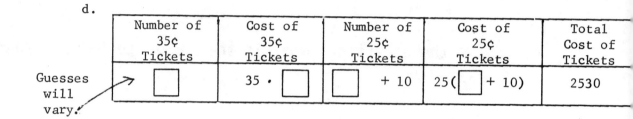

Guesses will vary.

Number of 35¢ Tickets	Cost of 35¢ Tickets	Number of 25¢ Tickets	Cost of 25¢ Tickets	Total Cost of Tickets
☐	35 · ☐	☐ + 10	25(☐ + 10)	2530

$$35N + 25(N + 10) = 2530$$

$$N = 38 \text{ tickets at } 35¢$$
$$N + 10 = 48 \text{ tickets at } 25¢$$

b. The sum of two numbers is 77.
   The larger number is three more than twice the smaller.
   What is the size of the smaller number?

c. Bill and Doug are bicycling toward each other on the same
   road. They are 36 miles apart. Suppose Bill averages
   17 miles per hour and Doug 13 miles per hour. How many
   hours will it take for them to meet one another?

d. Mary sold tickets to the carnival. Some rides were 25¢
   each. Others were 35¢ each. She sold ten more of the
   cheaper tickets. How many of each did she sell if she
   collected a total of $25.30 ?

TRANSLATING WORD PROBLEMS TO ALGEBRA

Mrs. Allen gave four word problems on Friday's test. The directions were

> Translate each problem to an algebraic equation. It is not necessary to solve the problems.

Jan's paper looked like the one below. Notice that the actual problems are not shown. Study her work carefully. In each case, write a problem that could have been the one she solved.

1. Let N be the smaller number.
   Then $3N$ is the larger number.
   $$N + 3N = 71$$

2. Let N be the number of nickels.
   Then $N + 5$ is the number of quarters.
   $$5N + 25(N + 5) = 425$$

3. Let N be the shortest side.
   Then $2N$ is the second side and
   $2N - 3$ is the third side.
   $$N + 2N + 2N - 3 = 87$$

4. Let N be Sarah's age.
   Then $N - 5$ is Barb's age and
   $4N$ is Doug's age.
   $$N + N - 5 + 4N = 55$$

PSM 82

Translating Word Problems To Algebra

Mathematics teaching objectives:

- Extract pertinent data and relationships from a word problem.
- Write an equation directly from a word problem.

Problem-solving skills pupils might use:

- Use mathematical symbols to describe situations.
- Solve a problem using a different method.
- Invent problems which can be solved by certain procedures.

Materials needed:

- None

Comments and suggestions:

- This lesson requires pupils to write an equation directly from word problems.
- This section provides an introduction to word problems. Pupils will need more practice from a textbook or other sources if they are to gain confidence in the translation process.
- At this stage it is better not to find a solution to the equations. If answers are stressed, the pupils will focus their attention on that rather than the translation process. Continue with this practice for a while when assigning word problems from other sources.
- Translation practice from textbooks should include both word phrases and word statements to their algebraic counterpart. Emphasize the difference between these two algebraic expressions.

Jan's Paper:

Answers will vary. Here are some possible responses.

1. The sum of two numbers is 71. One number is three times as large as the other. What are the numbers?

2. Jean has only quarters and nickels in her purse. The total value is $4.25. How many of each coin does she have if she has 5 more quarters than nickels?

3. The perimeter of a triangle is 87. The larger side is twice the shortest. The third side is three less than the longest side. Find the length of each side.

4. Barb is 5 years younger than Sarah. Doug is four times as old as Sarah. The sum of their ages is 55. How old is each person?

Translating Word Problems to Algebra (cont.)

Word problems can be solved by making guesses and using charts. Sometimes, however, it's possible to translate problems directly from words into algebra.

For each problem below, follow the steps that Jan used on the test. Be sure to begin each of them with the words, "Let N be ...." A solution for each problem is not necessary, only the translation.

1. One number is five times as large as another. Their sum is 78.
. What is the smaller number?

2. A bottle and a cork cost $1.05. The bottle costs a nickel more than the cork. How much does the cork cost?

3. The perimeter of a rectangle is 89 cm. The length is 8 cm more than the width. What is the width?

4. The sum of three consecutive numbers is 753. What is the smallest number?

5. A piece of wire 80 inches long is cut into two pieces. One piece is 5 inches more than twice the other. How long is the shorter piece?

6. Wanda has seven more dimes than nickels. The total value of the coins is $6.55. How many nickels does she have?

Translating Word Problem to Algebra

Answers:

The responses will vary. Pupils should become aware that different algebraic translations are possible for the same problem. Some possible translations are given.

1. Let N be the smaller number.   Let N be the larger number.

    Then 5N is the larger number.   or   Then $\frac{1}{5}$ N is the smaller number.

$$N + 5N = 78$$

$$N + \frac{1}{5} N = 78$$

2. Let N be the cost of the cork.

    Then $N + 5$ is the cost of the bottle.

$$N + N + 5 = 105$$

3. Let N be the width of the rectangle.

    Then $N + 8$ is the length.

$$N + N + 8 + N + N + 8 = 89 \quad \text{or:}$$

$$2N + 2(N + 8) = 89$$

4. Let N be the smallest number.

    Then $N + 1$ is the next number, and $N + 2$ is the largest number.

$$N + N + 1 + N + 2 = 753$$

5. Let N be the length of the shorter piece.

    Then $2N + 5$ is the length of the longer.

$$N + 2N + 5 = 80$$

6. Let N be the number of nickels.

    Then $N + 7$ is the number of dimes.

    5N is the value of the nickels, and $10(N + 7)$ is the value of the dimes.

$$5N + 10(N + 7) = 655$$

Algebra

VI. BINOMIALS

How often do your pupils make this mistake--
$$(x + 4)^2 = x^2 + 16$$

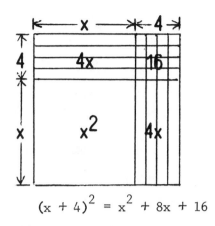

$$(x + 4)^2 = x^2 + 8x + 16$$

The intent of this four-lesson section on binomials is to provide pupils with a hands-on experience that might help eliminate the above common mistake.   The lessons use a visual model and the area concept, as shown in the diagram, for explaining the multiplication of two binomials.

The first two lessons use 10 by 10 squares, 10 by 1 strips, and 1 by 1 singles.   The final two lessons extend the procedure to x by x squares, x by 1 strips, and 1 by 1 singles.   Pupils should see that the algebraic calculations follow directly from the arithmetic ones.   By the end of this section, pupils should have discovered a method for multiplying binomials of the form

$$(x + a)(x + b) \quad \text{where a and b are positive.}$$

The class should explore the possibility of using the rectangular diagram procedure for products such as

$$(x - 2)(x - 3) \quad \text{or} \quad (x - 5)(x + 4).$$

Examples such as these are not dealt with in this section.   However, we recommend that pupils first try using squares, strips, and singles to show such products and then follow with a teacher-demonstration on the overhead.   The diagrams given below suggest a procedure that might be used.

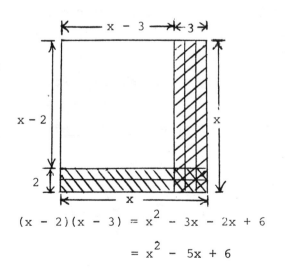

$$(x - 2)(x - 3) = x^2 - 3x - 2x + 6$$
$$= x^2 - 5x + 6$$

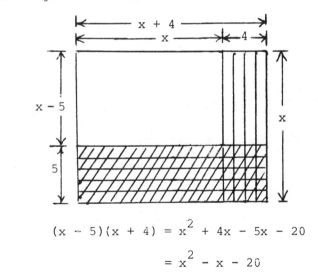

$$(x - 5)(x + 4) = x^2 + 4x - 5x - 20$$
$$= x^2 - x - 20$$

For more information on the visual approach to multiplication of binomials, see

Bidwell, James K. "A Physical Model For Fractoring Quadratic Polynomials," The Mathematics Teacher, Mar. 72, pp 201-205.

Miller, William. LABORATORY ACTIVITIES IN ALGEBRA. J. Weston Walch, Portland, Maine, 1974.

Rasmussen, Peter. MATHTILES, A CONCRETE APPROACH TO ARITHMETIC AND ALGEBRA. Key Curriculum Project, Berkeley,. CA, 1977.

Guess and check is an alternative approach for teaching multiplication of binomials. With the example $(x - 2)(x - 3)$, ask pupils to guess the product. Usually they will guess $x^2 - 5x - 6$, $x^2 + 6$, or $x^2 - 5x + 6$. To check their guesses, they can substitute several values for x and evaluate each expression. This "justification" technique has been used in previous PSM lessons and its application here would re-emphasize its importance.

Black masters for making the squares, strips, and singles are included. It is recommended that you run them off on tagboard or construction paper using three different colors. Some pupils soon will find they can do the problems using drawings or by visualizing the rectangles in their minds. Do not force the use of the pieces. Later, you might suggest that using the pieces may be easier than guessing and checking when finding the dimensions (or factoring) a rectangle with the area of $3x^2 + 7x + 2$.

These lessons should be used in sequence. Once the section is completed, more development and practice is needed.

Comments on variables -

On occasion, as in this section, variables are used in equations such as $(x + 4)^2 = x^2 + 8x + 16$ or $3(a + b) = 3a + 3b$. In these cases, any choice of the values for the variables x, a, and b will in each case produce true mathematical statements. This is not true for equations such as $n + 2 = 5$ or $x + y = 6$. In the sentence $n + 2 = 5$ there is only one value for n which makes the statement true. In the sentence $x + y = 6$, the choice of values for one of the variables is unrestricted, but the value, then, for the other variable is restricted to one value.

SQUARES

STRIPS

One half-sheet needed for each pupil.

SINGLES

wo sheets are needed for each class.

SM 82

RECTANGLES

Get 4 squares, 10 strips, and 15 singles.

Use the number of squares, strips, and singles shown. Make a rectangle. Make each side greater than one unit. Make a sketch of your rectangle. Record the dimensions of the rectangle. Record the total number of singles needed to cover the rectangle.

1. 3 strips,
   3 singles

2. 2 strips,
   2 singles

3. 3 strips,
   6 singles

dimensions: ___ , ___
total:_____

dimensions: ___ , ___
total:_____

dimensions:___ ,___
total:_____

4. Study these drawings for problem 1.

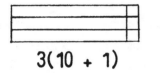

3(10 + 1)

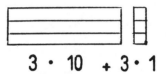

3 · 10 + 3 · 1

These drawings show this statement:

$$3 \cdot 11 = 3(10 + 1) = 30 + 3 = 33$$

a. Write a similar statement for prob. 2:

b. Write a similar statement for prob. 3:

Make rectangle drawings, record dimensions, and write a statement as in problem 4 for these:

5. 1 square,
   3 strips

6. 2 squares,
   2 strips

dimensions: ___ , ___
total:_____

Statement:

dimensions: ___ , ___
total: _____

Statement:

Rectangles

Mathematics teaching objectives:

  . Provide background for multiplying and factoring binomials.

  . Provide intuitive insight into the distributive property.

Problem-solving skills pupils _might_ use:

  . Make and use a drawing or model.

  . Guess and check.

  . Find other solutions.

Materials needed:

  . 10 cm by 10 cm squares, 10 cm by 1 cm strips, 1 cm by 1 cm squares
    need to be available for pupils if they need them.

Comments and suggestions:

  . The use of the squares, strips, and singles can be demonstrated at the
    overhead. Encourage pupils to use them on a few of the problems, but
    allow them, if they prefer, to work the problems through drawings.

  . Some pupils will use the distributive property in these exercises, but most
    will solve them by using the picture. Too much attention to the property
    could distract them from the main teaching objective for this lesson.

  . Encourage pupils to make their rectangles in problems 5 and 6 by grouping
    like shapes together. If this is done, the number statements for the
    rectangles may be easier for the pupils.

Answers:

1. ≡: , 3 by 11, 33, 3 x 11 = 3(10 + 1) = 30 + 3 = 33

2. ≡: , 2 by 11, 22, 2 x 11 = 2(10 + 1) = 20 + 2 = 22

3. ≡:: , 3 by 12, 36, 3 x 12 = 3(10 + 2) = 30 + 6 = 36

4. See answers 2 and 3 above.

5. ▢||| , 10 by 13, 130, 10 x 13 = 10(10 + 3) = 100 + 30 = 130

6. ▢▢|| , 10 by 22, 220, 10 x 22 = 10(20 + 2) = 200 + 20 = 220 <u>or</u>
    ▢▢ , 11 by 20, 220, 11 x 20 = (10 + 1)·20 = 200 + 20 = 220.

7. 4 squares,
    6 strips

    dimensions: ___ , ___
    total:_____
    Statement:

8. 8 strips,
    4 singles

    dimensions: ___ , ___
    total:_____
    Statement:

9. 6 strips,
    6 singles

    dimensions: ___ , ___
    total: _____
    Statement:

10. 6 strips,
    12 singles

    dimensions: ___ , ___
    total: _____
    Statement:

11. The rectangles in 8, 9, and 10 can be made in several different
    ways. Find another rectangle for each. Record the dimensions.
    Is the total different?

Rectangles

Answers:

7. , 20 by 23, 460, 20 · 23 = 20(20 + 3) = 400 + 60 = 460 <u>or</u>

, 10 by 46, 460, 10 · 46 = 10(40 + 6) = 400 + 60 = 460

8. , 4 by 21, 84, 4 x 21 = 4(20 + 1) = 80 + 4 = 84 <u>or</u>

2 by 42, 84, 2 · 42 = 2(40 + 2) = 80 + 4 = 84

9. , 6 by 11, 66, 6 · 11 = 6(10 + 1) = 60 + 6 = 66 <u>or</u>

, 3 by 22, 66, 3 · 22 = 3(20 + 2) = 60 + 6 = 66 <u>or</u>

, 2 by 33, 66, 2 · 33 = 2(30 + 3) = 60 + 6 = 66

10. , 6 by 12, 72, 6 · 12 = 6(10 + 2) = 60 + 12 = 72 <u>or</u>

, 3 by 24, 72, 3 · 24 = 3(20 + 4) = 60 + 12 = 72 <u>or</u>

, 2 by 36, 72, 2 · 36 = 2(30 + 6) = 60 + 12 = 72

11. See answers above for other possibilities.

MAKING RECTANGLES

Get 4 squares, 12 strips, and 15 singles.

For each problem, use the number of squares, strips, and singles shown. Make a rectangle. Place the square (or squares) in the lower left-hand corner. Draw a small sketch of the completed rectangle. Write the dimensions of the rectangle and the total number of singles needed to cover the rectangle.

1. 1 square,
   5 strips,
   4 singles

2. 1 square,
   6 strips,
   8 singles

3. 1 ☐, 9 ▯, 14 ▫

dimensions___,___   dimensions___,___   dimensions___,___
total ____   total ____   total ____

4. The diagram below shows the completed rectangle for
   problem 1. The mathematical statement can be written
   in the following forms:

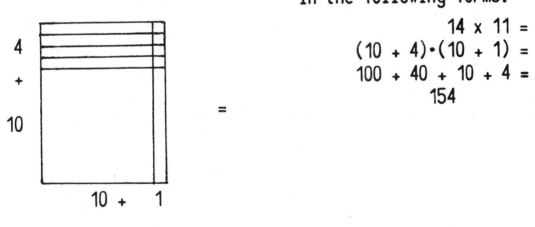

$$14 \times 11 =$$
$$(10 + 4) \cdot (10 + 1) =$$
$$100 + 40 + 10 + 4 =$$
$$154$$

Write a similar statement for problems 2 and 3:

Making Rectangles

Mathematics teaching objectives:

· Provide background for multiplying binomials.

· Provide a model useful for factoring.

Problem-solving skills pupils _might_ use:

· Make and use a drawing or model.

· Guess and check.

· Look for a pattern.

· Solve a problem using a different method.

Materials needed:

· 10 cm by 10 cm squares, 10 cm by 1 cm strips, 1 cm by 1 cm squares

Comments and suggestions:

· Many pupils can do this activity by drawing the squares and strips. You may want to have some of the squares and strips available for those who need them.

· Emphasize that the dimensions can be more easily read if the square is placed in the lower, left corner of the rectangle.

· Some pupils will make their rectangles with dimensions in a different order. Notice problem 4 on the pupil page and answer 1 below. Allow both possibilities.

· After problem 7, you might ask pupils if they have developed a plan or noticed a pattern. Some will have noticed they need to split up the number of strips into two groups, _a_ and _b_, whose "product" is the number of singles: $a + b$ = number of strips and $a \cdot b$ = number of singles. This plan works for problems 1-7, but not for 8-10. Why?

· Pupils will use different methods to find the total. Some will multiply; others will count the squares for the hundreds, strips for tens, etc.

Answers:

1.   , 11 by 14, 154

2.   , 12 by 14, 168

3.   , 12 by 17, 204

4. $12 \cdot 14 = (10 + 2)(10 + 4) = 100 + 40 + 20 + 8 = 168$

   $12 \cdot 17 = (10 + 2)(10 + 7) = 100 + 70 + 20 + 14 = 204$

ake rectangular drawings, record dimensions, and write a
tatement as in problem 4 for each of these:

. 1, 7, 6   (Abbreviation for   6. 1, 8, 15
            1□ , 7◻ , and 6◻)

dimensions: ___, ___   dimensions: ___, ___
total: _____   total: _____
Statement: Statement:

. 1, 8, 12   8. 2, 7, 6

dimensions: ___, ___   dimensions: ___, ___
total: _____   total: _____
Statement: Statement:

. 2, 11, 5   10. 4, 12, 5

dimensions: ___, ___   dimensions: ___, ___
total: _____   total: _____
Statement: Statement:

. Examine the statements from the problems above. How would these
  be finished?

    a. $(14 \cdot 13) = ($ $+$ $)($ $+$ $) = $ ___ $+$ ___ $+$ ___ $+$ ___ $=$

    b. $(15 \cdot 18) = ($ $+$ $)($ $+$ $) = $ ___ $+$ ___ $+$ ___ $+$ ___ $=$

Answers:

5. ▢||||||| , 11 by 16, 176, 11 · 16 = (10 + 1)(10 + 6) = 100 + 60 + 10 + 6

6. ▢||||| , 13 by 15, 195, 13 · 15 = (10 + 3)(10 + 5) = 100 + 50 + 30 + 1

7. ▢||||||| , 12 by 16, 192, 12 · 16 = (10 + 2)(10 + 6) = 100 + 60 + 20 + 1

8. ▢▢||| , 12 by 23, 276, 12 · 23 = (10 + 2)(20 + 3) = 200 + 30 + 40 + 6

9. ▢▢| , 15 by 21, 315, 15 · 21 = (10 + 5)(20 + 1) = 200 + 10 + 100 +

10. ▢▢▢▢||||| , 21 by 25, 525, 21 · 25 = (20 + 1)(20 + 5) = 400 + 100 + 20 +

11. a. 14 · 13 = (10 + 4)(10 + 3) = 100 + 30 + 40 + 12 = 182

  b. 15 · 18 = (10 + 5)(10 + 8) = 100 + 80 + 50 + 40 = 270

RECTANGLES IN ALGEBRA

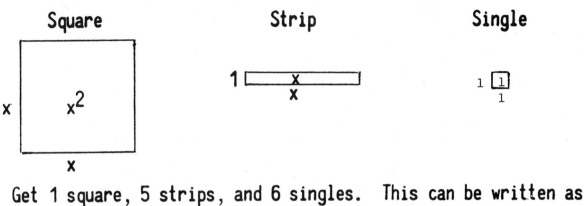

Square   Strip   Single

Get 1 square, 5 strips, and 6 singles. This can be written as $x^2 + 5x + 6$. Use the pieces to make a rectangle.

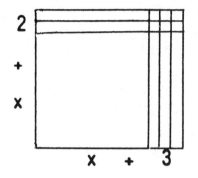

This shows the algebraic statement:

$$(x + 2)(x + 3) = x^2 + 3x + 2x + 6 = x^2 + 5x + 6.$$

1. Use the number of squares, strips, and singles shown. Make a rectangle. Write an algebraic statement describing the rectangle.

    a. 1 square, 7 strips, 6 singles

    b. 1, 7, 12

    c. 1, 6, 9

    d. 1, 9, 18

    e. 2, 5, 2

    f. 4, 8, 3

Rectangles In Algebra

Mathematics teaching objectives:

- Provide background for multiplying binomials.
- Use a model for factoring trinomials.
- Factor trinomials $x^2 + bx + c$ where c is constant or where b is constant

Problem-solving skills pupils _might_ use:

- Make and use a drawing or model.
- Guess and check.
- Look for a pattern.
- Use mathematical symbols to describe situations.

Materials needed:

- Squares, strips, and singles.

Comments and suggestions:

- Some pupils may find they can draw the rectangles rather than use the pi
- It is important to emphasize that the algebraic explanation is modeled t same as the arithmetic explanation. For example:

$$(10 + 2)(10 + 3) = 100 + 30 + 20 + 6 = 156$$
$$\text{and} \quad (x + 2)(x + 3) = x^2 + 3x + 2x + 6 = x^2 + 5x + 6$$

Later, the vertical format can be shown:

$$
\begin{array}{r}
10 + 3 \\
\underline{10 + 2} \\
20 + 6 \\
\underline{100 \ + 30 \qquad} \\
100 \ + 50 + 6
\end{array}
\qquad
\begin{array}{r}
x + 3 \\
\underline{x + 2} \\
2x + 6 \\
\underline{x^2 + 3x \qquad} \\
x^2 + 5x + 6
\end{array}
$$

- Problems 2 and 3 suggest a pattern for factoring trinomials with a leadir coefficient of 1, namely, two numbers, a and b, are needed such that $a + b$ = number of strips and $a \times b$ = number of singles.

- Problem 4 can provide a good visual way of illustrating the concepts involved in "completing the square."

Answers: Each answer could be commuted; that is, $(x + 1)(x + 6) = (x + 6)(x +$

1. a. , $(x + 1)(x + 6) = x^2 + 6x + 1x + 6 = x^2 + 7x + 6$

   b. , $(x + 3)(x + 4) = x^2 + 4x + 3x + 12 = x^2 + 7x + 12$

   c. , $(x + 3)(x + 3) = x^2 + 3x + 3x + 9 = x^2 + 6x + 9$

   d. , $(x + 3)(x + 6) = x^2 + 6x + 3x + 18 = x^2 + 9x + 18$

   e. , $(x + 2)(2x + 1) = 2x^2 + 1x + 4x + 2 = 2x^2 + 5x + 2$

   f. , $(2x + 1)(2x + 3) = 4x^2 + 6x + 2x + 3 = 4x^2 + 8x + 3$

-174-

Rectangles in Algebra (cont.)

2. Use the number of squares, strips, and singles shown. Make a
    rectangle. Write an algebraic statement to describe the rectangle.

    a. 1, 10, 24
    b. 1, 11, 24
    c. 1, 14, 24
    d. 1, 25, 24

    e. 1, 7, 12
    f. 1, 8 12
    g. 1, 13, 12

3. Use the number of squares, strips, and singles shown. Make a
    rectangle. Write an algebraic statement to describe the rectangle.

    a. 1, 10, 9
    b. 1, 10, 16
    c. 1, 10, 21
    d. 1, 10, 24
    e. 1, 10, 25

4. One part is missing. Use what is given to find the missing
    part so the resulting rectangle is a square. Write an
    algebraic statement to describe the square.

    a. 1, ___, 9
    b. 1, ___, 4
    c. 1, ___, 36
    d. 1, 10, ___
    e. 1, 2, ___
    f. 1, 8, ___

Answers:

2. a. ▦ , $(x + 4)(x + 6) = x^2 + 6x + 4x + 24 = x^2 + 10x + 24$

    b. ▦ , $(x + 3)(x + 8) = x^2 + 8x + 3x + 24 = x^2 + 11x + 24$

    c. ▦ , $(x + 2)(x + 12) = x^2 + 12x + 2x + 24 = x^2 + 14x +$

    d. ▦ , $(x + 1)(x + 24) = x^2 + 24x + 1x + 24 = x^2 + 25x$

    e. ▦ , $(x + 3)(x + 4) = x^2 + 4x + 3x + 12 = x^2 + 7x + 12$

    f. ▦ , $(x + 2)(x + 6) = x^2 + 6x + 2x + 12 = x^2 + 8x + 12$

    g. ▦ , $(x + 1)(x + 12) = x^2 + 12x + 1x + 12 = x^2 + 13x +$

3. a. ▦ , $(x + 1)(x + 9) = x^2 + 9x + 1x + 9 = x^2 + 10x + 9$

    b. ▦ , $(x + 2)(x + 8) = x^2 + 8x + 2x + 16 = x^2 + 10x + 16$

    c. ▦ , $(x + 3)(x + 7) = x^2 + 7x + 3x + 21 = x^2 + 10x + 21$

    d. ▦ , $(x + 4)(x + 6) = x^2 + 6x + 4x + 24 = x^2 + 10x + 24$

    e. ▦ , $(x + 5)(x + 5) = x^2 + 5x + 5x + 25 = x^2 + 10x + 25$

4. a. 6, ▦ , $(x + 3)(x + 3) = x^2 + 3x + 3x + 9 = x^2 + 6x + 9$

    b. 4, ▦ , $(x + 2)(x + 2) = x^2 + 2x + 2x + 4 = x^2 + 4x + 4$

    c. 12, ▦ , $(x + 6)(x + 6) = x^2 + 6x + 6x + 36 = x^2 + 12x + 36$

    d. 25, ▦ , , $(x + 5)(x + 5) = x^2 + 5x + 5x + 25 = x^2 + 10x + 25$

    e. 1, ▦ , $(x + 1)(x + 1) = x^2 + 1x + 1x + 1 = x^2 + 2x + 1$

    f. 16, ▦ , $(x + 4)(x + 4) = x^2 + 4x + 4x + 16 = x^2 + 8x + 16$

THE MISSING PARTS

One of the three parts of the algebraic statement is shown. Work through each problem completely before going on to the next.

A	B	C

Example: $(x + 4)(x + 2) = x^2 + 2x + 4x + 8 = x^2 + 6x + 8$

        A

1. $(x + 7)(x + 3)$ _____   _____

2. _____ $x^2 + 2x + 9x + 18$ _____

3. _____   _____ $x^2 + 8x + 15$

4. $(x + 5)(x + 6)$ _____   _____

5. _____ $x^2 + 4x + 6x + 24$ _____

6. _____   _____ $x^2 + 5x + 4$

7. $(x + 15)(x + 2)$ _____   _____

8. _____ $x^2 + 7x + 7x + 49$ _____

9. _____   _____ $x^2 + 16x + 48$

10. $(x + 5)(2x + 3)$ _____   _____

11. _____ $2x^2 + 2x + 3x + 3$ _____

12. _____   _____ $3x^2 + 7x + 2$

The Missing Parts:

Mathematics teaching objectives:

- Multiply binomials.
- Factor trinomials.

Problem-solving skills pupils might use:

- Make and use a drawing or model.
- Look for a pattern.
- Guess and check.
- Visualize an object from its drawing.

Materials needed:

- Squares, strips, and singles (optional)

Comments and suggestions:

- Some pupils may just draw the rectangles instead of using the pieces.

- Emphasize the drawing or model and how it matches up for each of the three ways of labeling the rectangle. Use the example to show part A (factor or dimension form), part B (partial product form), and part C (final product form).

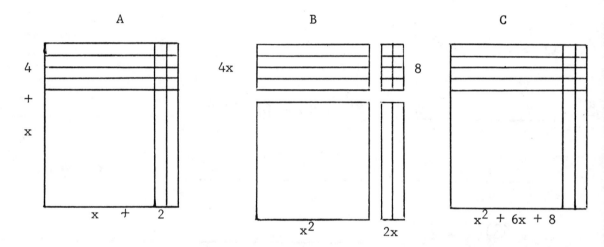

- By this time, pupils should recognize the sum-product pattern. For a trinomial with a leading coefficient of 1, two numbers, a and b, are needed such that $a + b$ = number of strips and $a \cdot b$ = number of singles.

- Spend some class time discussing the distributive property and the role of variable in equations such as $3(x + 5) = 3x + 15$, $(x + 2)(x + 3) = x^2 + 5x + 6$, and $x + 2 = 5$. See page 158.

Answers: Some answers could be commuted, that is, $(x + 3)(x + 5) = (x + 5)(x + 3)$

1. $x^2 + 3x + 7x + 21$, $x^2 + 10x + 21$

2. $(x + 9)(x + 2)$, $x^2 + 11x + 18$

3. $(x + 3)(x + 5)$, $x^2 + 5x + 3x + 15$

4. $x^2 + 6x + 5x + 30$, $x^2 + 11x + 30$

5. $(x + 6)(x + 4)$, $x^2 + 10x + 24$

6. $(x + 1)(x + 4)$, $x^2 + 4x + 1x + 4$

7. $x^2 + 2x + 15x + 30$, $x^2 + 17x + 3$

8. $(x + 7)(x + 7)$, $x^2 + 14x + 49$

9. $(x + 4)(x + 12)$, $x^2 + 12x + 4x +$

10. $2x^2 + 3x + 10x + 15$, $2x^2 + 13x +$

11. $(2x + 3)(x + 1)$, $2x^2 + 5x + 3$

12. $(3x + 1)(x + 2)$, $3x^2 + 6x + 1x +$

Algebra

VII.   GRAPHS AND EQUATIONS

Prior to taking first-year algebra, most pupils have been exposed to both graphing and the geometry of quadrilaterals in separate introductory units. This section builds on this background together with the understanding they have gained in equation solving. The intention is to give an intuitive development of equations and their graphs. These lessons relate ordered pairs, graphs, and equations in a variety of ways. In some cases, pupils are asked to determine an equation from a set of ordered pairs derived from a graph. In others, the equation is given and a graph is determined.

Many ideas come up in the lessons. The equations and graphs of horizontal and vertical lines are mixed in the problems. Some of the graphs and equations relate to curves. Vocabulary is enlarged and reinforced through the use of words like origin, coordinate, y-axis, parallel, and quadrant. These lessons set the stage for more formal work with slope and y-intercept to be taught later in the year.

The activities in this section are arranged in a logical and teachable sequence from beginning to end and they can be taught at any time before work with slope and y-intercept, provided pupils have the ability to plot points, read graphs, and find pairs of points that satisfy simple equations.

The only materials needed for this section are graph paper and rulers.

 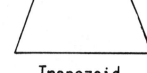

quare   Rectangle   Parallelogram   Trapezoid

A quadrilateral has four sides. They can be identified by the coordinates of the vertices (corner points). Match each set of points in the chart with the most appropriate quadrilateral pictured above.

	Corner Points	Quadrilateral
a.	(2,3) (2,8) (10,3) (10,8)	
b.	(⁻2,⁻3) (⁻2,3) (4,3) (4,⁻3)	
c.	(⁻1,⁻1) (3,3) (3,⁻5) (7,⁻1)	
d.	(0,⁻6) (0,8) (⁻4,⁻4) (⁻4,6)	
e.	(2,2) (3,7) (11,6) (12,11)	

In each case, three vertices of a quadrilateral are given. Determine the coordinates of a fourth vertex.

	Quadrilateral	Vertices
a.	Square	(⁻1,⁻1) (⁻2,5) (4,6) (,)
b.	Rectangle	(0,0) (2,⁻3) (⁻6,⁻4) (,)
c.	Parallelogram	(4,2) (6,8) (12,11) (,)
d.	Trapezoid	(⁻6,⁻3) (⁻3,6) (6,6) (,)

Plotting Points

Mathematics teaching objectives:

. Provide practice in plotting points and reading coordinates of points.

. Apply geometry concepts.

Problem-solving skills pupils <u>might</u> use:

. Make and use a drawing.

. Visualize an object from its description.

. Guess and check.

. Find another answer.

Materials needed:

. cm grid paper (see page 163)

. Straight edge or ruler (transparent one preferable)

Comments and suggestions:

. Review the definitions of square, rectangle, parallelogram, and trapezoid. Discuss reasons why a square is also classified as a rectangle and parallelogram and a rectangle as a parallelogram.

. Review the analytical geometry terms <u>axes</u>, <u>origin</u>, <u>quadrants</u>, <u>coordinates</u> <u>of a point</u> (<u>x-coordinate</u>; <u>y-coordinate</u>), and the plotting of points given the coordinates and vice versa. Remind pupils that a reversing of the order of the coordinates names a different point.

. Work problems la and lb with the class before letting them proceed on their own individually.

. Insist that graph paper be used to solve and/or record their solutions.

Answers:

1. a. rectangle   b. square   c. square   d. trapezoid   e. parallelogram

2. a. (5,0)   b. ($^-$4,$^-$7)   c. (10,5); (14,17); or ($^-$2,$^-$1)

d. Infinitely many solutions. One solution set: points with coordinates of the form (x,$^-$3) except points (3,$^-$3), ($^-$6,$^-$3) and ($^-$15,$^-$3). All these points are on the same straight line.

Two other solution sets are points on the lines y = 3x − 12 and $y = \frac{3}{4} x + \frac{33}{4}$.

3. a and b. (4,7) and (8,1) or (8,$^-$1) and ($^-$4,$^-$7). If the two original points are opposite corners of a diagonal, the other two corners of the square are (3,2) and ($^-$3,$^-$2)

c,d,e. Infinitely many solutions. Have pupils check each other's solution.

4. a,b,c. Three possibilities exist. (13,5), (7,13), or ($^-$3,$^-$1).

3. Plot the points $(2,^-3)$ and $(^-2,3)$. These are two vertices of a quadrilateral. Locate and give the coordinates of two other vertices so the quadrilateral is

   a. a square.
   b. a square different than a.
   c. a rectangle that is not a square.
   d. a parallelogram that is not a rectangle.
   e. a parallelogram other than b, c, or d.
   f. a trapezoid.

4. Plot the points $(5,2)$, $(2,6)$, and $(10,9)$. These are three vertices of a parallelogram.

   a. Locate and give the coordinates of a fourth vertex.

   b. Locate and give the coordinates of another point which could be the fourth vertex.

   c. How many different points could be a fourth vertex? Give the coordinates of any other vertex you find.

DRAWING GRAPHS

Draw these graphs.

1. a. The graph is a straight line.
   b. It goes through the origin.
   c. It goes through the point ($^-$3,5).

2. a. The graph is a circle.
   b. The origin is the center of the circle.
   c. It passes through the point ($^-$10,0).

3. a. The graph is a straight line.
   b. It is parallel to the x-axis.
   c. The y-coordinate is always $^-$3.

4. a. The graph is a curved line.
   b. The graph is only in the first quadrant.
   c. It passes through: (2,12) (8,3) (6,4) (12,2).

5. a. The graph is a straight line.
   b. It slopes sharply up from lower left to upper right.
   c. It passes through ($^-$2,$^-$6) and (2,6).

6. a. The graph is a straight line.
   b. The x-coordinate is always equal to the y-coordinate.
   c. It passes through the origin.

7. a. The graph is a straight line.
   b. It slopes up from left to right at a 45° angle.
   c. It passes through a point with coordinates (0,$^-$1).

8. a. The graph is a straight line.
   b. It is parallel to the y-axis.
   c. The point (4,0) is one unit to the left of the graph.

Drawing Graphs

Mathematics teaching objectives:

- Build vocabulary related to graphing.
- Apply knowledge of graphing and straight lines.

Problem-solving skills pupils <u>might</u> use:

- Satisfy one condition at a time.
- Eliminate extraneous information.
- Make a graph.

Materials needed:

- Graph paper
- Straight edge or ruler

Comments and suggestions:

- Encourage pupils to search the problem for a clue they can use immediately such as "It goes through (2,5)." Then they can work with the other clues.
- Ask pupils to identify the problems that can be solved even if one condition is crossed off. For example, problem 5 could have condition <u>b</u> eliminated and problem 6 could have condition <u>c</u> eliminated.
- Some pupils might want to use a protractor for problem 7.
- Have pupils compare the graphs for problems 6 and 7.

Answers:

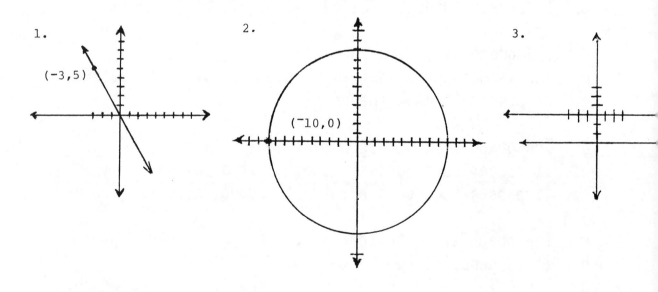

9.   a.   The graph is a straight line.
      b.   It does not go through the origin.
      c.   It does go through both (0,3) and (5,0).

10.   a.   The graph is a straight line.
      b.   The y-coordinate is always 0.
      c.   It intersects the y-axis at the origin.

11.   a.   The graph is a straight line.
      b.   The graph does not slope up or down.
      c.   It passes through (0,⁻6).

12.   a.   The graph is a straight line.
      b.   It slopes down from left to right.
      c.   The sum of the coordinates is 6.

13.   a.   The graph is a straight line.
      b.   It goes through (7,2).
      c.   The coordinates of each point have the same sign
           (positive,positive) or (negative,negative).

14.   a.   The graph is a straight line.
      b.   It slopes down from left to right.
      c.   The y-coordinate is the opposite of the x-coordinate.

15.   a.   The graph curves down from the left, then curves up
           to the right.
      b.   It passes through the origin.
      c.   The y-coordinate is the square of the x-coordinate.

Drawing Graphs

Answers:

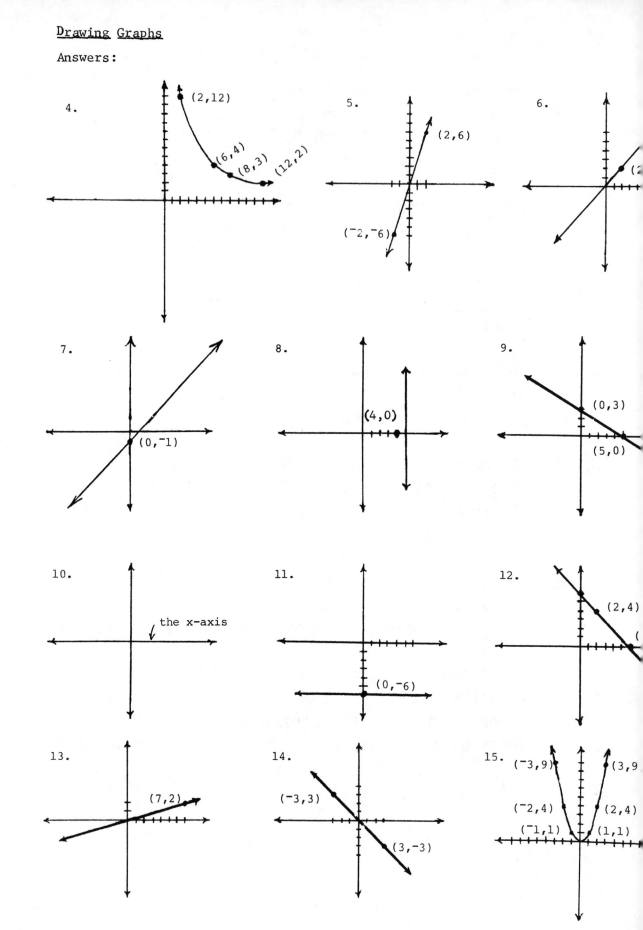

POINTS TO EQUATIONS

The sets of ordered pairs are coordinates of points:

    a. Plot the ordered pairs.
    b. Sketch the graph suggested by the points.
    c. Describe the graph. Use words.
    d. Write three more ordered pairs that would be on the graph.
    e. Study the ordered pairs for a pattern (or common property). Use words to describe the pattern.
    f. Write an algebraic equation that describes the pattern in part e.

1. (1,2) (5,6) (⁻1,0) (⁻3,⁻2) (0,1) (3,4)

2. (4,8) (4,⁻3) (4,0) (4,4) (4,⁻7)

3. (⁻4,16) (⁻3,9) (⁻2,4) (⁻1,1) (0,0) (1,1) (2,4) (3,9) (4,16)

4. (10,0) (9,1) (8,2) (5,5) (3,7) (0,10) (⁻4,14)

5. (⁻2,⁻2) (0,⁻2) (⁻4,⁻2) (6,⁻2) (1,⁻2)

Mathematics teaching objectives:

. Describe graphs in words.

. Write equations for sets of ordered pairs.

Problem-solving skills pupils <u>might</u> use:

. Look for a pattern.

. Find likenesses and differences and make comparisons.

. Use mathematical symbols to describe situations.

. Make and use a graph.

Materials needed:

. Graph paper

. Straight edge

Comments and suggestions:

. Review the meaning of the term "ordered pairs."

. Have pupils re-read the descriptions given in the problems for the preceeding lesson ("Drawing Graphs") for suggestions on ways to describe their graphs.

. Remind pupils to state anything special about straight-line graphs. Do they cross the origin? Slope up? Are they horizontal?

. Pupils will have difficulties with problems 2 and 5. Notice that the answers are given as $x = 4 + 0 \cdot y$ and $y = {}^-2 + 0 \cdot x$, rather than $x = 4$ and $y = {}^-2$. This is to emphasize the other coordinate is still part of the graph. Many pupils see $x = 4$ as having no y values. However, emphasize that equations such as $x = 4$ and $y = {}^-2$ are legitimate equations for lines.

Answers:

1.

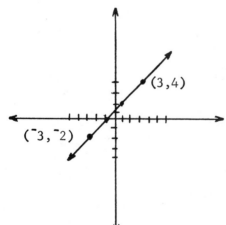

2.

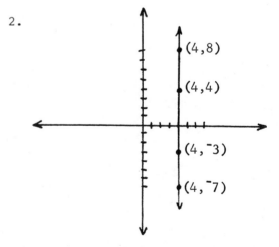

c. A straight line sloping up left to right.
d. Answers vary.
e. The y-coordinate is one more than the x-coordinate.
f. $y = x + 1$

c. A straight line 4 units to the right of the y-axis and paralle to it. The y-coordinate is any real number.
d. Answers vary.
e. The x-coordinate is always 4.
f. $x = 4 + 0 \cdot y$.

4.

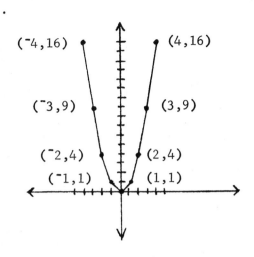

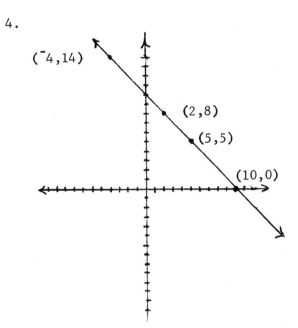

c. The graph is a curved line
    passing through the origin.
    (You may wish to introduce
    the term "parabola.")
d. Answers vary.
e. The y-coordinate is the
    square of the x-coordinate.
f. $y = x^2$

c. A straight line sloping down
d. Answers vary.
e. Their sum is 10.
f. $x + y = 10$

5.

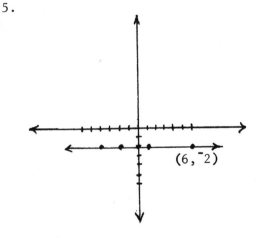

c. Straight line 2 units below and
       parallel to the x-axis
d. Answers vary.
e. The y-coordinate is always ⁻2.
           x is any real number.
f. $y = {}^-2 + 0 \cdot x$

GRAPHS TO EQUATIONS

For each graph shown below:

    a. Describe the graph. Use words.
    b. Write six ordered pairs that lie on the graph.
    c. Study the ordered pairs for a pattern (or property).
       Use words to describe the pattern.
    d. Write an algebraic equation that describes the
       pattern in part <u>c</u>.

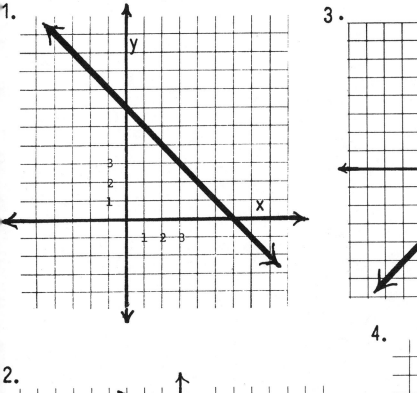

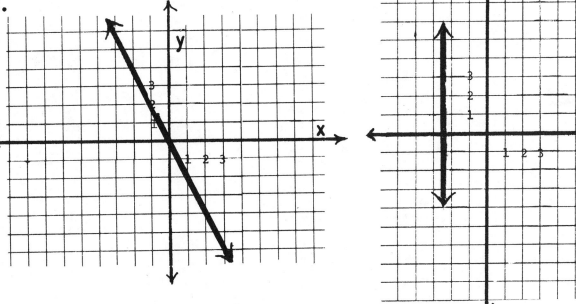

Graphs To Equations

Mathematics teaching objectives:

    . Read graphs.

    . Write equations for graphs.

Problem-solving skills pupils might use:

    . Look for a pattern.

    . Find likenesses and differences and make comparisons.

    . Use mathematical symbols to describe situations.

    . Use a graph.

Materials needed:

    . None

Comments and suggestions:

    . In many graph lessons, pupils move in this sequence:

$$\text{equations} \longrightarrow \text{ordered pairs} \longrightarrow \text{graphs.}$$

    Here, the reverse is true:

$$\text{graphs} \longrightarrow \text{ordered pairs} \longrightarrow \text{equations.}$$

    . As a challenge, provide graphs whose equations are more difficult to determine, e.g., $x + 2y = 10$.

Answers:

1. a. A straight line sloping down left to right, intersecting the y-axis at $(0,6)$ and the x-axis at $(6,0)$.
     b. Answers vary.
     c. In each, the sum of the coordinates is 6.
     d. $x + y = 6$

2. a. A straight line sloping up left to right at $45°$ going through the origin.
     b. Answers vary.
     c. In each, the coordinates are equal.
     d. $y = x$

3. a. A straight line sloping down sharply from left to right and going through the origin.
     b. Answers vary.
     c. In each, the x-coordinate is $\frac{1}{2}$ the negative of the y-coordinate (or the y-coordinate is 2 times the negative of the x-coordinate).
     d. $x = \frac{^-1}{2}y$ or $y = {}^-2x$.

4. a. A vertical line $2\frac{1}{2}$ units left of the y-axis.
     b. Answers vary.
     c. The x-coordinate is always $^-2\frac{1}{2}$.
     d. $x = {}^-2\frac{1}{2} + 0 \cdot y$.

Graphs to Equations (cont.)

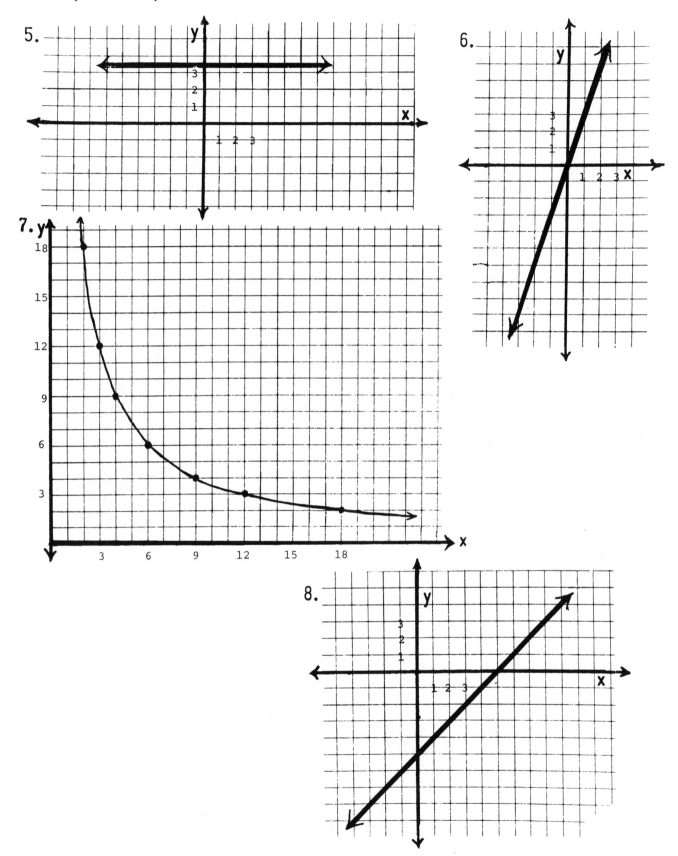

Answers:

5. a. A horizontal line, $3\frac{1}{2}$ units above the x-axis.

    b. Answers vary.

    c. In each, the y-coordinate is always $3\frac{1}{2}$.

    d. $y = 3\frac{1}{2} + 0 \cdot x$.

6. a. A straight line sloping sharply from left to right and going through the origin.

    b. Answers vary.

    c. In each, the y-coordinate is 3 times the x-coordinate.

    d. $y = 3x$.

7. a. A curve in quadrant I which approaches both the y-axis and the x-axis.

    b. Answers vary.

    c. In each, the product of the coordinates is 36.

    d. $xy = 36$, where x and y both are positive.

8. a. A straight line sloping up intersecting the y-axis at (0,5) and x-axis at (5,0).

    b. Answers vary.

    c. In each, the difference between the x-coordinate and the y-coordinate is 5.

    d. $x - y = 5$

WORDS TO EQUATIONS

For each description below:

a. Write four ordered pairs that fit the description.
b. Plot the ordered pairs.
c. Sketch the graph suggested by the points.
d. From the graph write three more ordered pairs that lie
   on the graph. Check the numbers with the word description.
e. Write an algebraic equation for the word description.

1. The sum of the two numbers is eight.

2. The difference between the two numbers is two.

3. The quotient of the two numbers is three.

4. The first number is always zero.

5. The second number is always negative four.

6. The second number is two more than the square of the
    first number.

7. The second number is the opposite (or inverse) of the
    first number.

8. Twice the first number added to the second number is twelve.

Words To Equations

Mathematics teaching objectives:

- Graph relationships.
- Write equations to fit word descriptions.

Problem-solving skills pupils <u>might</u> use:

- Use mathematical symbols to describe situations.
- Make reasonable estimates.
- Look for a pattern.
- Make and use a graph.

Materials needed:

- Graph paper and straight edge

Comments and suggestions:

- This lesson gives pupils experience in moving from
  statements $\longrightarrow$ ordered pairs $\longrightarrow$ graphs $\longrightarrow$ equations.
- Problems 2 and 3 have two possible answers, depending on the order
  pupils choose for x and y.
- Notice that in problem 3 the point (0,0) is not allowed. This problem
  provides the opportunity to review the division properties of zero.
- Problem 6 offers two important instructional opportunities. Many
  pupils will connect the points suggested by the problem with a broken
  rather than a curved line. Also, they will find it difficult to find
  numbers which check exactly with the word description.

Answers: Answers for <u>a</u> and <u>d</u> of each problem will vary.

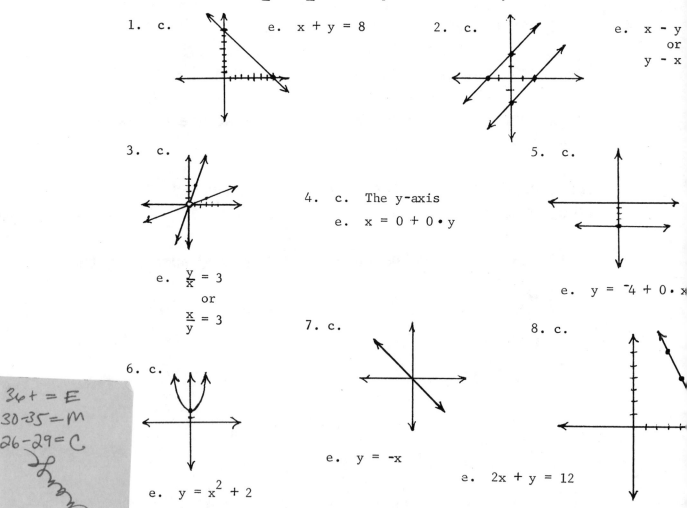

1. c.   e. x + y = 8   2. c.   e. x - y
                                                            or
                                                           y - x

3. c.   4. c. The y-axis

                                 e. x = 0 + 0 · y   5. c.

    e. $\frac{y}{x} = 3$
          or
        $\frac{x}{y} = 3$
                             7. c.   8. c.

                                                            e. y = ⁻4 + 0 · x

6. c.

                                 e. y = -x

                                          e. 2x + y = 12

    e. $y = x^2 + 2$

36+ = E
30-35 = M
26-29 = C

EQUATIONS TO POINTS TO GRAPHS

1. a. Circle the ordered pairs that satisfy this equation: $2x + y = 8$.

    (3,2)   (⁻1,10)   (4,4)   (6,⁻4)   (0,6)

    (8,⁻8)   (⁻5,2)   (0,8)   (1,6)   (4,2)

    b. Plot all the ordered pairs in problem 1.

    c. Circle the points that satisfied the equation. What do
    you notice?

2. a. Circle the ordered pairs that satisfy this equation: $y = 3x + ⁻4$.

    (3,5)   (⁻2,⁻10)   (0,⁻4)   (2,2)   (5,11)

    (⁻1,7)   (0,0)   (1,⁻1)   (⁻2,⁻2)   (⁻3,⁻13)

    b. Plot all the ordered pairs in problem 2.

    c. Circle the points that satisfied the equation. What do
    you notice?

3. a. The equation is $x - y = 4$. One coordinate of several ordered
    pairs is given. Find the other coordinate.

    (4,)   (6,)   (2,)   (, 3)   (,⁻1)

    (0,)   (⁻2,)   (,⁻3) $(7\frac{1}{2}, $)   (,$1\frac{1}{2}$)

    b. Plot the ordered pairs. Draw the graph.

4. a. The equation is $y = \frac{1}{2}x + 3$.   The x- coordinate is given.

    Find the y- coordinate for each.

    (0,)   (⁻6,)   (⁻2,)   (6,)

    (2,)   (8,)   (⁻4,)   (4,)

    b. Plot the ordered pairs. Draw the graph.

Equations To Points To Graphs

Mathematics teaching objectives:

- Determine ordered pairs that satisfy equations.
- Relate equations, points, and graphs.

Problem-solving skills pupils <u>might</u> use:

- Find likenesses and differences and make comparisons.
- Look for a pattern.
- Make a graph.

Materials needed:

- Graph paper
- Straight edge

Comments and suggestions:

- To satisfy an equation means that the values suggested for the variables <u>x</u> and <u>y</u> make the equation a true statement.

- This lesson takes pupils from equations to ordered pairs satisfying the equations, to graphs. It is the opposite in development from the lesson <u>Graphs To Equations</u>.

- Ask pupils to compare the equations for straight and curved lines. Ask, "What equation properties allow you to accurately predict whether the graph will be a straight line?" The equations for curved lines from previous lessons are $xy = 36$ and $y = x^2 + 2$.

- An overhead transparency of the page entitled General Equations for Lines (see page 205) should be used when the lesson is discussed-- probably the next class session. The listing of the equations includes those graphed for the lesson and leads to the general linear equation forms. Have pupils answer the questions on the transparency. The form $y = mx + b$ will be developed further in the next section.

Answers:

1.a. $(3,2)(^-1,10)(6,^-4)(8,^-8)$
$(0,8)(1,6)$

b.

2.a. $(3,5)(^-2,^-10)(0,^-4)(2,2)(5,11)$
$(1,^-1)(^-3,^-13)$

b.

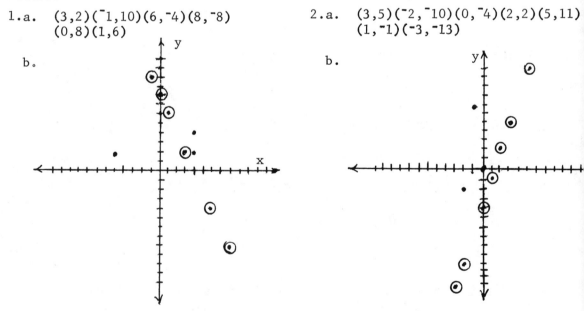

c. They lie on a straight line.　　　　c. They lie on a straight line.

5. The equation is x + y = ‾5.

   a. Find ordered pairs that fit the following conditions:
     . The x-coordinate is zero. (,)
     . The y-coordinate is one less than the x-coordinate. (,)
     . The y-coordinate is zero. (,)
     . The x-coordinate is positive. (,)
     . The x-coordinate is four times the y-coordinate. (,)

   b. Plot the points in part <u>a</u>. Draw the graph of x + y = ‾5.

6. The equation is y = 2x + ‾6.

   a. Find ordered pairs that fit the following conditions:
     . The x-coordinate is zero. (,)
     . Both coordinates are negative. (,)
     . The x-coordinate is positive and the y-coordinate is
      negative. (,)
     . The y-coordinate is one-half the x-coordinate. (,)
     . The y-coordinate is 0. (,)
     . Both coordinates are positive. (,)

   b. Plot the points in part <u>a</u>. Draw the graph of y = 2x + ‾6.

7. a. The equation is 2x + 3y = 24. Find six ordered pairs that
     fit the equation. Arrange them in this table.

x						
y						

   b. Plot the points. Draw the graph of 2x + 3y = 24.

8. a. The equation is $y = \frac{2}{3}x$. Find six ordered
     pairs that fit the equation. Write them
     in this T-table.

x	y

   b. Plot the points. Draw the graph of $y = \frac{2}{3}x$.

9. Write an equation whose graph will be a curved
   rather than a straight line. Check your equation
   by making a table of values, plotting points and
   then drawing the graph of your equation.

Equations To Points To Graphs

Answers:

3.a. $(4,0)(6,2)(2,^-2)(7,3)(3,^-1)(0,^-4)$
$(^-2,^-6)(1,^-3)(7\frac{1}{2},3\frac{1}{2})(5\frac{1}{2},1\frac{1}{2})$

4.a. $(0,3)(^-6,0)(^-2,2)(6,6)(2,4)(8,7)$
$(^-4,1)(4,5)$

b.

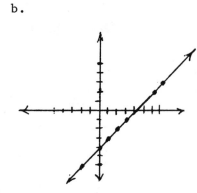

b.

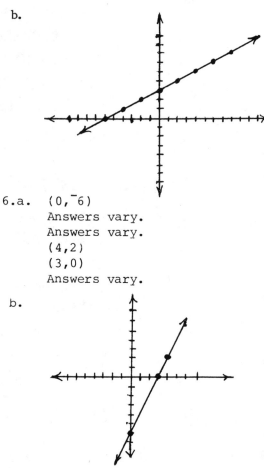

5.a. $(0,^-5)$
$(^-2,^-3)$
$(^-5,0)$
Answers vary.
$(^-4,^-1)$

6.a. $(0,^-6)$
Answers vary.
Answers vary.
$(4,2)$
$(3,0)$
Answers vary.

b.

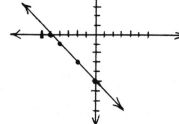

b.

7.a. Answers vary.

b.

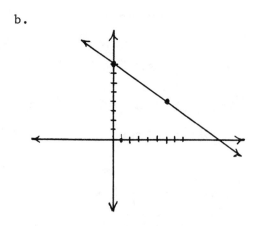

8.a. Answers vary.

b.

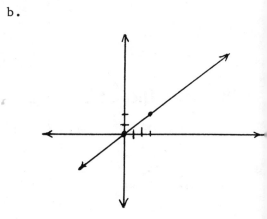

9. Answers will vary. Remind pupils: Equations $xy = 36$ and $y = x^2 + 2$ from previous lessons have graphs which are curved lines.

GENERAL EQUATIONS FOR LINES
A Lesson Summary

$$2x + y = 8$$

$$x - y = 4$$

$$x + y = {}^-5$$

$$2x + 3y = 24$$

$$x = 6$$

$$y = 5$$

$$y = 3x - 4$$

$$y = \frac{1}{2}x + 3$$

$$y = 2x - 6$$

$$y = \frac{2}{3}x$$

$$y = 10$$

These are linear equations of the form

$$ax + by = c$$

These are linear equations of the form

$$y = mx + b$$

What are the values of a, b, and c in each of the above equations?

What are the values of m and b in each of the above equations?

Algebra

VIII. GRAPH INVESTIGATIONS

This section begins with the determination of the slope of a line from its graph in relatively easy cases and ends with a lesson on the determination of a line of best fit from data collected from the real world. The lessons in between, along with supplementary work from other sources, explore ways for determining

- the equation of a line from its slope and y-intercept, from its slope and a point on the line, and from two points on the line.
- whether or not a table of values describes a straight line.
- whether or not, from its equation, the graph is a straight line.

Slope usually is introduced in first year algebra. It grows in importance as a mathematical topic until calculus where rates of change are studied and applied in depth. Although slope has long been recognized as an important topic in mathematics, it remains difficult to teach. Pupils often reverse the change in y with the change in x, use the coordinates of points incorrectly to find the slope of a line joining two points or forget to solve an equation for y before using the coefficient of x for the slope. The use of a "variable" for a "constant" in $y = mx + b$ causes confusion.

One way to help pupils understand slope is to have them discover, through applying problem solving skills, how to determine the slope from a graph and how the slope of a line relates to its equation. Through discovery, they have a chance to make the ideas of slope their own.

This section includes determining the slope of a line from its graph, determining the slope of a line joining two points, relating the y-intercept and slope to an equation and line of best fit. Care is taken to ease pupils into the use of m and b as general values for constants.

Before attempting these activities, pupils must have skills in plotting points, reading graphs, and graphing equations by picking appropriate points. The lessons are arranged in a logical order; however, only the problem solving or discovery parts are provided here. In order for the lessons to be successful, teachers will need to:

- provide many applications and practice problems for each lesson through textbook or worksheet assignments.
- take an active part in guiding the discovery lesson during class--facilitate discussion, check on progress, summarize, etc.

Materials needed are graph paper and rulers.

SLOPE

Some of the lines below are quite steep. Others are not. You'll note that below each line there's a ratio. This ratio, called the <u>slope</u> <u>of</u> <u>the</u> <u>line</u>, is used to show how steep it is.

Can you discover how to determine the slope of a line? Study the graphs carefully before working the exercises.

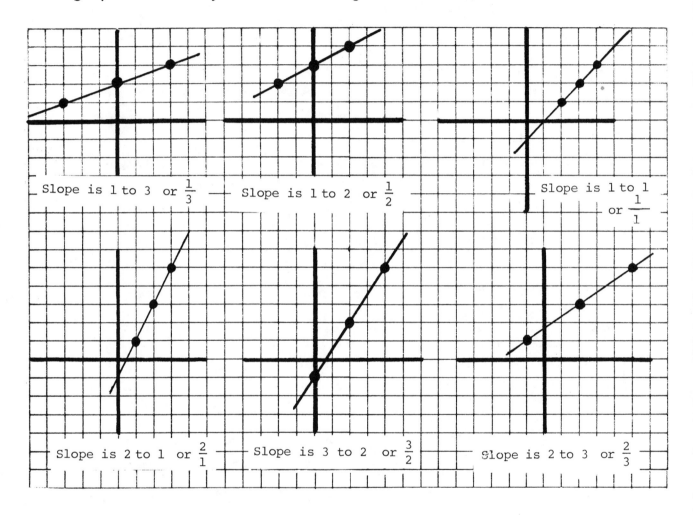

Slope is 1 to 3 or $\frac{1}{3}$

Slope is 1 to 2 or $\frac{1}{2}$

Slope is 1 to 1 or $\frac{1}{1}$

Slope is 2 to 1 or $\frac{2}{1}$

Slope is 3 to 2 or $\frac{3}{2}$

Slope is 2 to 3 or $\frac{2}{3}$

<u>Slope</u>

Mathematics teaching objectives:

- Introduce slope.
- Discover how to determine a slope from a graph of a straight line.
- Practice finding the slope of a line given 2 points on the line.

Problem-solving skills pupils <u>might</u> use:

- Make conjectures based upon data.
- Guess and check.
- Look for a pattern.

Materials needed:

- Graph paper (see page 236)

Comments and suggestions:

- Do this lesson in class. Let pupils try the exercise on their own or with a partner until they have tried problem 1. Then any assistance should be as nondirective as you can manage.
- Discuss with the pupils their answers to 2, 3, and 7 as they are working, preferably on a one-to-one basis or in small groups.
- <u>A</u> <u>book</u> <u>assignment</u> <u>or</u> <u>worksheet</u> <u>will</u> <u>be</u> <u>needed</u> <u>to</u> <u>provide</u> <u>the</u> <u>extra</u> <u>practice</u> <u>needed</u> <u>in</u> <u>determining</u> <u>slopes</u>.
- Delay the formal use of y = mx + b until the lesson on <u>Slope -- Intercept Equation</u> (page 227).

Answers:

1. a. 4   b. ⁻1   c. 1   d. ⁻2   e. 1   f. $\frac{1}{3}$

2. One possibility: If the line climbs left to right, it is positive; if it falls left to right it is negative.

3. Answers will vary. One possibility is to compare the amount of change in the y-direction with the amount of change in the x-direction.

Slope (cont.)

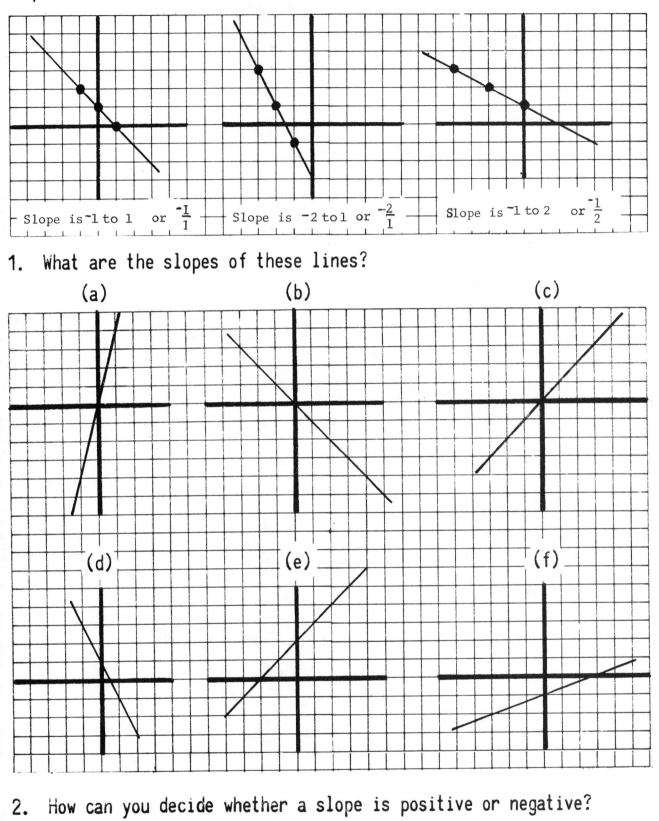

Slope is ⁻1 to 1   or $\frac{-1}{1}$ Slope is ⁻2 to 1 or $\frac{-2}{1}$ Slope is ⁻1 to 2   or $\frac{-1}{2}$

1. What are the slopes of these lines?

(a)   (b)   (c)

(d)   (e)   (f)

2. How can you decide whether a slope is positive or negative?

3. Describe how to find the slope of a line.

Slope (cont.)

4. In each case, plot the two points. Then determine the slope of the line joining the two points.

    a. (1,2) (4,5)   c. (0,0) (5,3)   e. (5,0) (0,1)
    b. (2,1) (⁻1,4)   d. (0,3) (1,6)   f. (⁻2,⁻2) (0,2)

5. Study the graph sketches below. Note that grids have not been drawn. Nevertheless, you should still be able to find the slope of each line.

    Keep track of the method you're using. It should help you later.

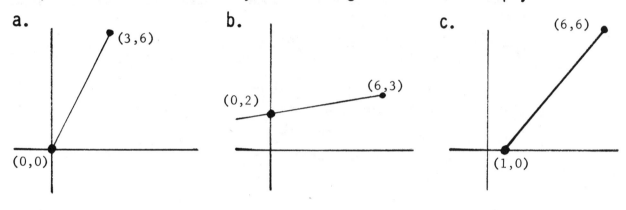

a. (3,6) (0,0)

b. (0,2) (6,3)

c. (6,6) (1,0)

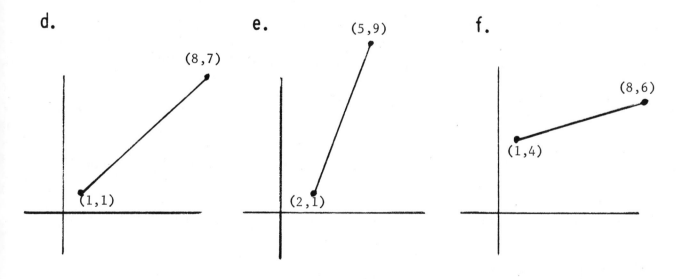

d. (8,7) (1,1)

e. (5,9) (2,1)

f. (8,6) (1,4)

Answers:

4. a. 1   b. ⁻1   c. $\frac{3}{5}$ d. 3   e. $\frac{-1}{5}$ f. 2

5. a. 2   b. $\frac{1}{6}$ c. $\frac{6}{5}$ d. $\frac{6}{7}$ e. $\frac{8}{3}$ f. $\frac{2}{7}$

6. a. $\frac{10}{7}$ b. $\frac{9}{6}$ c. $\frac{8}{5}$ d. $\frac{5}{3}$ e. 1   f. $\frac{5}{6}$ g. $\frac{d-}{c-}$

7. Answers will vary. However, after listening to some of the pupils'
    methods, be sure they realize the slope (positive or negative) of a
    line joining two points (a,b) and (c,d) can be determined by finding

    $\frac{d-b}{c-a}$ or $\frac{b-d}{a-c}$. This can be related to the change in y divided by the

    change in x. Pupils often have trouble keeping the coordinates straight.
    Sometimes a sketch helps.

    Additional practice with this concept will be needed. Provide practice
    in graphing a line if given a point and the slope of the line.

    For example: Given (3,4) and a slope of $\frac{-2}{3}$. Start at (3,4) then find
    a second point by using the change
    in the y-coordinate as ⁻2 and the
    change in the x-coordinate as 3.

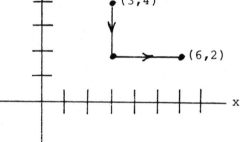

6. In each case, determine the slope of the line joining the two points. Do this without drawing the graphs.

   a. (0,0) (7,10)

   b. (1,1) (7,10)

   c. (2,2) (7,10)

   d. (2,2) (5,7)

   e. (2,3) (7,8)

   f. (3,1) (9,6)

   g. (a,b) (c,d)

7. The slopes of all the lines in problems 5 and 6 are positive. Does your method work for lines that have negative slopes?

   Try your method using the points below. Check the slopes by actually plotting the points. If they don't check, you need to revise your method.

   a. (6,0) (0,5)

   b. (⁻2,4) (7,1)

HOW FAST DOES A BABY GROW?

Mr. Russell presented some data to his class showing the weekly growth of his small son.

He indicated that a baby's growth during the first several weeks usually forms a linear pattern.

If this information is plotted, we see that the points do suggest a straight line. Mr. Russell called the line which is drawn "a line of best fit."

Birth	
1 Week	$8\frac{1}{2}$ lbs.
2 Weeks	$9\frac{1}{4}$ lbs.
3 Weeks	$9\frac{3}{4}$ lbs.
4 Weeks	$9\frac{3}{4}$ lbs.
5 Weeks	$10\frac{1}{2}$ lbs.
6 Weeks	11 lbs.
7 Weeks	$11\frac{1}{2}$ lbs.
8 Weeks	$11\frac{3}{4}$ lbs.

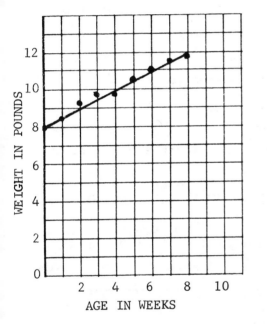

Many questions can be asked about this situation. For example –

. How long do you suppose a growth pattern like this will last?

. If it did continue according to the pattern, how much will the "baby" weigh at the end of 1 year? 2 years? 14 years?

We could answer these questions by drawing the graph on a large piece of paper. By extending the line we could then read the weights which correspond to 1 year, 2 years, etc. It will be much simpler, however, if we can find an equation which fits the graph. Then we can use the equation to determine the various weights.

Slope - Y-Intercept

Mathematics teaching objectives:

. Predict the slope and y-intercept from an equation.

○ Write equations for lines given the slope and y-intercept.

. Use an equation to make predictions.

Problem-solving skills pupils might use:

. Look for a pattern.

. Simplify the problem.

. Make conjectures based upon data.

. Make reasonable estimates.

Materials needed:

. Graph paper (see page 236)

Comments and suggestions:

. The baby problem provides the motivation for a careful examination of equations and their graphs. Emphasize that the simpler problems can give clues to the more complicated baby problem.

. Review the method for graphing a line when its equation is given (see page 201).

. Note that a table of values accompanied the graph of the growth for Mr. Russell's son. Suggest a table of values be made for the equations in problem 1 before points are plotted.

. More practice in relating slopes and y-intercepts to equations will be necessary.

. A master for an overhead transparency (see page 225) gives the answers to problems 1-5. The organization of the answers clearly shows the meanings for the constants $\underline{m}$ and $\underline{b}$ in $y = mx + b$. Use the transparency when discussing this lesson.

Answers:

1. a. 2   b. ⁻2   c. 3   d. $\frac{1}{2}$

2. a. Answers will vary.   b. ⁻3

3. a. slope = 2, y-intercept = 3   b. slope = ⁻2, y-intercept = ⁻1
    c. slope = 3, y-intercept = 2   d. slope = $\frac{1}{2}$, y-intercept = ⁻2

4. a. Answers will vary.   b. slope = ⁻3, y-intercept = 4

5. a. $y = 4x + 3$ b. $y = \frac{1}{2}x + 8$

6. a. $y = \frac{1}{2}x + 8$ where $\underline{x}$ is the age in weeks and $\underline{y}$ is the weight in pounds.

    b. 34 lbs., 60 lbs., 372 lbs. An analysis of some specific data reveals that this growth pattern does not continue much beyond the 3rd month.

    c. Answers will vary.

Slope — Y-Intercept (cont.)

These exercises will help you discover a method that can be used on problems such as this one. We will return to the questions regarding the baby's weight later.

1. First, draw the graph of the equation. Then find its slope.

   a. $y = 2x$

   b. $y = -2x$

   c. $y = 3x$

   d. $y = \frac{1}{2}x$

2. Study both the equations and slopes in problem 1.

   a. Predict the slope of the graph of $y = -3x$.

   b. Check your prediction by drawing the graph of $y = -3x$. What is its slope?

3. The y-intercept is the point of intersection of the graph and the y-axis.

   First, draw the graph of each equation.
   Next, find its slope.
   Finally, find the y-intercept.

   a. $y = 2x + 3$

   b. $y = -2x - 1$

   c. $y = 3x + 2$

   d. $y = \frac{1}{2}x - 2$

4. Study the equations, slopes, and y-intercepts in problem 3.

   a. Predict the slope and y-intercept of the graph of $y = -3x + 4$.

   b. Check your predictions by drawing the graph of $y = -3x + 4$. What is its slope? What is its y-intercept?

5. Write the equations of lines with

    a. a slope of 4 and a y-intercept of 3.

    b. a slope of $\frac{1}{2}$ and a y-intercept of 8.

    c. a slope of m and a y-intercept of b.

6. Examine the graph of the baby's weight given at the beginning.

    a. What is the equation for the graph?

    b. If the pattern continued, what would the baby's weight be at the end of 1 year?   2 years?   14 years?

    c. Within what age limits do you think the equation could be used to make reasonable predictions?

SLOPE — Y-INTERCEPT

Answers

Equation	Slope	y-intercept
1. a. $y = 2x$	2	0
b. $y = {}^-2x$	$^-2$	0
c. $y = 3x$	3	0
d. $y = \frac{1}{2}x$	$\frac{1}{2}$	0
2. $y = {}^-3x$	$^-3$	0
3. a. $y = 2x + 3$	2	3
b. $y = {}^-2x - 1$	$^-2$	$^-1$
c. $y = 3x + 2$	3	2
d. $y = \frac{1}{2}x - 2$	$\frac{1}{2}$	$^-2$
4. $y = {}^-3x + 4$	$^-3$	4
5. a. $y = 4x + 3$	4	3
b. $y = \frac{1}{2}x + 8$	$\frac{1}{2}$	8
c. $y = mx + b$	m	b

SLOPE -- INTERCEPT EQUATION

Which problem at the right
do you think would be the
more difficult to solve?
Why?

> A line goes through (2,5). It has
> a slope of 2. What is its
> y-intercept?

> A line goes through (50,41). It
> has a slope of $\frac{1}{2}$. What is its
> y-intercept?

1. Let's solve the easier problem first by making a graph.

   a. What is the y-intercept of the line described in the
      first box?

   b. What is the equation of the first line? Use the
      slope-intercept form: $y = mx + b$

      slope   y-intercept

2. The problem in the first box can be solved easily by graphing.
   A graphing solution to the second problem is not as easy.
   Let's find a way that does not rely on drawing a graph.
   You know these things about the second graph:

   . The line is straight and slopes from left to right.
     Therefore, its equation is of the form $y = mx + b$.

   . The slope is $\frac{1}{2}$. Therefore, the equation looks like
     $y = \frac{1}{2}x + b$.

   . The line goes through (50,41). Therefore, (50,41) must
     satisfy the final equation.

   a. Will either of the equations below be the correct one?
      Try (50,41) in each of them to see.

      $$y = \frac{1}{2}x + 10 \qquad\qquad y = \frac{1}{2}x + 20$$

   b. What is the equation of the line?

   c. What is its y-intercept?

Slope -- Intercept Equation

Mathematics teaching objectives:

- Write the equation of a line given its slope and a point on the line.
- Determine whether or not the graph described by a table of values is a straight line.
- Find an efficient way for finding the equation for a line given two points.

Problem-solving skills pupils might use:

- Guess and check.
- Make use of a graph.
- Solve an easier but related problem.
- Solve a problem using a different method.

Materials needed:

- Graph paper (see page 236).

Comments and suggestions:

- Discuss why the second introductory problem is more difficult than the first one.
- The more difficult equation is used to motivate a search for an easier way. Emphasize that frequently simpler problems give insight into solving more difficult ones.
- Review "Equations To Points To Graphs" along with the transparency for the answers (pages 201-205). The presentation of the answers should help pupils see the meanings for m and b in y = mx + b.
- Review a procedure for solving problem 1. See comments for the answer to 1a.
- Remind pupils that slope is $\dfrac{\text{change in y}}{\text{change in x}}$.
- Before letting pupils proceed independently on problems 4 and 5, remind them that
  - the slope between any two points on a straight line is always the same.
  - the graph is a straight line if the slope between any two points is the same.

Answers:

See page 230.

Slope - Intercept Equation (cont.)

3. Write the equations of lines which satisfy these conditions:

   a. Slope of 3 and goes through (7,11).

   b. Slope of ⁻2 and goes through (⁻8,1).

   c. Slope of $\frac{1}{3}$ and goes through (12,4).

   d. Slope of $\frac{9}{5}$ and goes through (10,50).

Find a better way than graphing for working problem 4 and 5.

4. In each part, decide whether the points are on the same line. If they are, find the slope and write the equation.

a.

x	y
2	3
3	5
4	7
5	9

b.

x	y
0	3
⁻1	4
⁻2	5
⁻3	6

c.

x	y
4	0
6	2
8	6
10	10

d.

x	y
10	12
9	14
8	16
7	18

5. It was a pleasant day in the resort city of Costa Del Sol in Spain. The sun was out; people were swimming. The temperature was 25⁰. On the same day in London it was cool and drizzly. People wore their sweathers. The temperature was 59⁰. By studying the weather report we see that temperatures have been recorded in both Fahrenheit and Celsius.

	Celsius (x value)	Fahrenheit (y value)	
Helsinki	10	50	Overcast
London	15	59	Showers
New York	20	68	Cloudy
Costa Del Sol	25	77	Sunny
Cairo	30	86	Sunny

   a. Explain why these points all will lie on the same straight line.

   b. The temperature in Mexico City is 43⁰ Celsius. What is the corresponding temperature in Fahrenheit?

Slope -- Intercept Equation

Answers:

1. a. 1

   Comment: Have pupils graph (2,5) and use what they know about slope to find another point. They can then draw the line and see where it intersects the y-axis.

   b. $y = 2x + 1$

2. a. No, but the next guess for the y-intercept should be between 10 and 20.

   b. $y = \frac{1}{2}x + 16$

   c. 16

3. a. $y = 3x + ^-10$
   b. $y = ^-2x + ^-15$
   c. $y = \frac{1}{3}x$
   d. $y = \frac{9}{5}x +$

4. a. $y = 2x + ^-1$
   b. $y = ^-x + 3$
   c. Not a straight line. The slope is 1 between the first two points and 2 between the next two point

   d. $y = ^-2x + 32$

5. a. The slope of the line joining any two points is always $\frac{9}{5}$.

   b. $F = \frac{9}{5}C + 32$ so $43^{\circ}C = 109.4^{\circ}F$

LINE OF BEST FIT

1. Paul made up the following height-weight formula:

   $$W = 10H - 500$$ H is the person's height in inches.
   W is the weight in pounds.

   a. Substitute in some heights to see how accurate the formula is.

   b. Do you think Paul's formula is reasonable? Explain.

2. Other members of the class decided to try to determine their own height-weight formula. First, they collected information on some people in their neighborhood. The first number of each pair represents the height in inches; the second number represents the weight in pounds.

(50,67)	(60,99)	(63,119)	(74,179)
(70,149)	(73,169)	(55,70)	(72,180)
(51,59)	(54,80)	(62,135)	(52,63)
(48,55)	(64,121)	(65,130)	(61,100)
(46,50)	(61,119)	(67,141)	(66,136)
(64,130)	(54,86)	(56,85)	(56,91)

   a. Plot the points suggested by these ordered pairs. Let the height-axis be the horizontal axis.

   b. As with most information collected from the "real world," it does not fit an exact mathematical pattern. However, the plotted points do seem to cluster around a straight line. Draw in the line which you think fits best.

   c. Write the formula for your line of best fit.

Line Of Best Fit

Mathematics teaching objectives:
- Graph "live" data.
- Determine graph for the line of best fit.
- Determine the least and greatest number in a range of values.

Problem-solving skills pupils might use:
- Make reasonable estimates.
- Look for patterns.
- Make predictions based upon data.

Materials needed:
- Special graph paper (see page 235).
- Transparent ruler

Comments and suggestions:
- The major purpose of this lesson is to demonstrate the approximate nature of mathematics in making predictions in the "world of reality." The orde pairs in problem 2 are actual heights and weights of real people.
- Ask pupils why the graph paper (page 235) has different scales on each a Have pupils locate the ordered pair that
    - shows the height of the shortest person; the tallest person.
    - the weight of the "lightest" person; the heaviest person.
- Discuss the concept of range of values.
- Encourage pupils to exercise care in plotting the points. Plotting becon difficult and tedious when the numbers are large and the scales are diffe ent on each axis. Possibly pupils could work in pairs.
- Elicit from pupils the best way to draw the "line of best fit." A trans-
- parent ruler would help.
- Allow pupils to discuss in small groups ways to determine the slope of th line, the y-intercept and equation for the line.
- Compare the equations pupils have for their lines. The equations will vary.
- Elicit from the pupils why the equations vary.

Answers:
1. a. Answers will vary.
   b. Answers will vary.
2. See page 234.

d. According to your formula,

. what is the weight of a person 5 feet tall?

. what is the weight of a 3-year old who is 38 in. ?

. what is the height of a professional football player who weighs 270 pounds?

e. Determine your own weight by means of the formula. Does the formula give a fairly accurate prediction in this case?

f. For what heights, if any, does the formula seem to be reasonably accurate?

2. a & b

c. Answers will vary. One possibility
    is $y = \frac{8}{3}x - 43$.

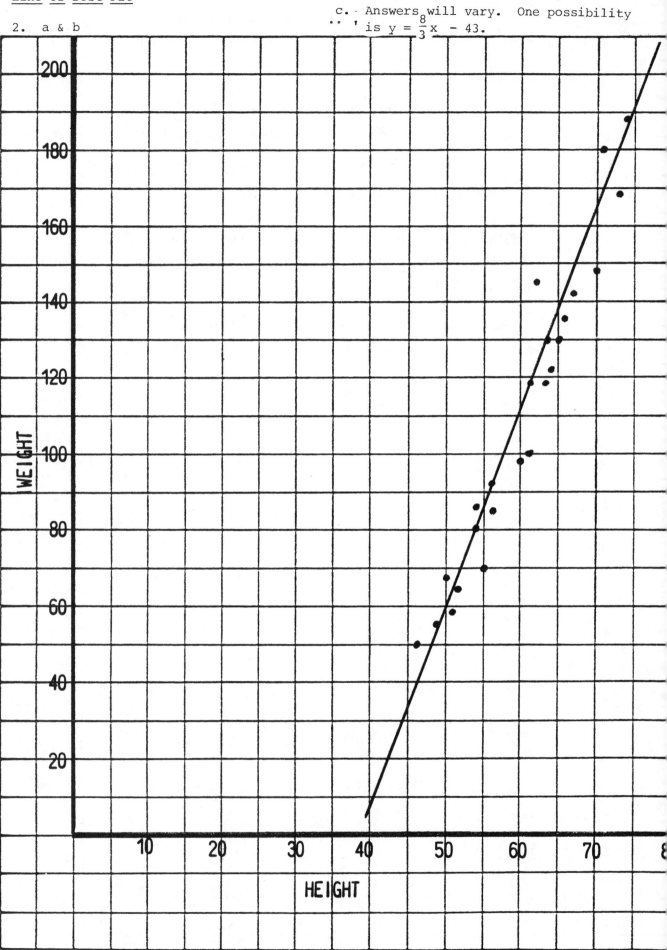

RELATIONSHIP BETWEEN A PERSON'S HEIGHT AND WEIGHT

WEIGHT

180

160

140

120

100

80

60

40

20

10   20   30   40   50   60   70   80

HEIGHT

PSM 82

Algebra

IX.   SYSTEMS OF LINEAR EQUATIONS

IX. SYSTEMS OF LINEAR EQUATIONS

This section presents three initial exploratory lessons on systems of equations. Explicit use is made of the problem-solving skills, making a systematic list, finding another answer, eliminating possibilities, solving a problem using a different method, making graphs, guessing and checking, and making generalizations based upon data. The remaining pages in this section are teacher suggestions for lessons that will be needed before pupils have the necessary algebraic skills for efficiently solving systems of linear equations. The spirit of the suggestions is to provide ideas for maintaining aspects of a problem-solving approach throughout the rest of the unit.

Before using this section, pupils should be able to graph equations, recognize that $y = mx + b$ and $ax + by = c$ are general equations for straight lines, and use slope and y-intercept concepts when graphing lines. This section is an extension of ideas presented in the sections Graphs And Equations and Graph Investigations.

As a way of introducing systems of equations, the first lesson uses two relatively easy number puzzles, each with infinitely many answer-pairs. However, only one answer-pair is a solution common to both puzzles. The pupils' method of solution involves using guessing and checking and making a systematic list. In most of the problems the corresponding equations accompany the puzzles.

In the second lesson, pupils experience difficulty solving pairs of equations using the methods of the first lesson and are directed to use graphing. Systems are encountered which have one, no, or infinitely many solutions. More practice with the ideas presented in this lesson will be needed before going on to the final lesson given in this section. Most algebra textbooks have materials that can be used.

In the third lesson, the limitation of the graphical method is demonstrated and pupils are guided in their search for a more efficient method. The approach used is summarized on page 250. The approach is different and should be studied carefully before it is used in class.

A general comment -

An equation such as $x = 3$ has several interpretations. In one context it means

- one answer for a number puzzle
- a point on a number line
- an equation for a line in a coordinate plane.

All these interpretations need to be made explicit when teaching this section.

Comment on terminology -

The terminology used in this section is consistent with usage in the following paragraphs.

An equation such as 2x + y = 5 has two variables and there are many values for x and y which make the equation a true statement. Any pair of these values is a solution for this equation. Usually, the solutions are expressed as ordered pairs, e.g., (2,1) or (3,⁻1). Each ordered pair satisfied the equation.

Two equations, each using the same symbols for variables, are called a system of equations. A solution to a system of linear equations in two variables is an ordered pair of numbers which satisfy both equations. A system of linear equations may have no solution, one or infinitely many solutions.

I'M THINKING OF TWO NUMBERS

1. I'm thinking of two numbers.
   If you add them, the answer is 7.
   What are the numbers?

   a. How is this number puzzle different from others you
      have solved?
   b. Give 5 possible solutions.
   c. How many solutions are possible?

2. I'm thinking of two numbers.
   If you subtract the two numbers, the answer is 3.
   What are the numbers?

   a. How is this puzzle different from others you have solved?
   b. Give 5 possible solutions.
   c. How many solutions are possible?
   d. Is there a common solution that works for both puzzles
      given above? If so, what is it?

3. I'm thinking of two numbers.
   Their sum is 20. $(x + y = 20)$
   Their difference is 6. $(x - y = 6)$

   a. Give 5 possible solutions for the "$x + y = 20$" condition.
   b. Give 5 possible solutions for the "$x - y = 6$" condition.
   c. Find a solution that fits both conditions.

<u>I'm</u> <u>Thinking</u> <u>Of</u> <u>Two</u> <u>Numbers</u>

Mathematics teaching objectives:

- Introduce systems of equations (two linear equations).

- Introduce one solution, no solution, infinitely many solutions for two linear equations.

Problem-solving skills pupils <u>might</u> use:

- Guess and check.

- Make a systematic list or table.

- Eliminate possibilities.

- Look for patterns.

Materials needed:

- Graph paper (see page 236)

Comments and suggestions:

- Much of this activity needs to be teacher-directed with class discussion on each problem. Let pupils work independently after problem 3.

- You will need to check pupil progress on problem 6 which has infinitely many solutions and number 8 which has no solution.

- Observe the solution attempts for problem 5. Most pupils will work it the same as the others. However, someone might choose to substitute 2y for x in x + y = 15 and then solve for y and x. If this happens, defer commenting about this originality until the lesson summary.

- In the lesson summary, emphasize that there can be one, no, or infinitely many solutions for a system of linear equations.

Answers:

1. a. Not enough information to determine the numbers thought of.

   b. Answers will vary.   c. Infinite number

2. a,b,c. Same as above.   d. Yes; 5 and 2

3. a & b. Answers will vary.   c. 13 and 7

4. a & b. Answers will vary.   c. 12 and ⁻2

5. 10 and 5 (see comments and suggestions)

6. Any pair of numbers that differ by 2. Discussion should bring out responses as to why this pair of equations has more than one solution.

7. x = 2, y = 1

8. Not possible. Again, discussion is necessary.

I'm Thinking of Two Numbers (cont.)

4.   I'm thinking of two numbers.
        Their sum is 10. $(x + y = 10)$
        If you add the first to twice the second,
          the result is 8. $(x + 2y = 8)$

    a. Give 5 possible solutions for the first condition.
    b. Give 5 possible solutions for the second condition.
    c. Find a solution that fits both conditions.

5.   I'm thinking of two numbers.
        The first is twice as large as the second. $(x = 2y)$
        Their sum is 15. $(x + y = 15)$

    What are the numbers?

6.   I'm thinking of two numbers.
        The first minus the second is 2. $(x - y = 2)$
        Twice the first minus twice the second is 4. $(2x - 2y = 4)$

    What are the numbers?

7. Write a number puzzle that fits these two equations.
    Then solve the puzzle by finding the two numbers.

        $x + 3y = 5$
        $3x + y = 7$

8. Write a number puzzle that fits these two equations.
    Then solve the puzzle by finding the two numbers.

        $x + 2y = 7$
        $x + 2y = 11$

SOLVING SYSTEMS OF LINEAR EQUATIONS

$$x + y = 13$$
$$x - y = 7$$

EASY

$$x + 3y = 10$$
$$3x - 2y = {}^-14$$

DIFFICULT

1. Find the solution to the "easy" system of equations.

2. There is a solution to the second system of equations. But
    it is not easy to find by a guess-and-check procedure.
    Spend a few minutes to see if you can find the solution.

3. A graph can be helpful in finding the solution to the second
    system.

    a. Draw the graph of $x + 3y = 10$.
        How many solutions are possible for this equation?

    b. On the same grid, draw the graph of $3x - 2y = {}^-14$.
        How many solutions are possible for this equation?

    c. What ordered pair satisfies both equations?
        Check your solution to see if it fits both equations.

4. Use graphing to find the common solution to each system of
    equations.

    a. $x + y = {}^-4$
          $^-x + y = 2$

    b. $^-2x + y = 4$
          $4x - y = {}^-2$

    c. $x - y = 6$
          $2x + y = 0$

    d. $x + y = {}^-2$
          $0 \cdot x + y = {}^-6$

    e. $x + 0 \cdot y = 3$
          $0 \cdot x + y = {}^-2$

    f. $2x + y = 2$
          $y = {}^-2x + 6$

    g. $4x + 2y = 6$
          $2x = {}^-y + 3$

Solving Systems Of Linear Equations

Mathematics teaching objectives:

- Introduce graphing to solve systems of linear equations.
- Relate number of solutions to parallel, intersecting, or coinciding lines.

Problem-solving skills pupils _might_ use:

- Guess and check.
- Make and use a graph.
- Make generalizations based upon data.
- Solve a problem using a different method.

Materials needed:

- Graph paper

Comments and suggestions:

- Use the entire lesson. It may take two class periods.
- The more difficult problem is used to motivate the graphing method for solving pairs of linear equations. Time spent on problem 2 will convince pupils that there must be a better way. Many pupils will have difficulty graphing the difficult equations.
- Anticipate some difficulty with zero coefficients in problem 4, parts d and e. Review the concept using a transparency for page 225. In the next lesson, pupils can use the concept that a straight line graph can be described by many different equations, e.g., $2x + y = 3$, $4x + 2y = 6$, $6x + 3y = 9$, etc. These equations are called _equivalent equations_.
- Supplement this lesson with related materials from the class text or worksheets. Pupils will need additional practice in recognizing equivalent equations and changing a given linear equation to an equivalent one. They also need to recognize inconsistent pairs of equations. However, the practice should be limited to those equations which are in the $ax + by = c$ form.

Answers:

1. $x = 10$, $y = 3$

2. It is doubtful that many pupils will guess the solution.

3. a. Infinite   b. Infinite   c. $(^-2, 4)$

4. a. $(^-3, ^-1)$ b. $(1, 6)$ c. $(2, ^-4)$

    d. $(4, ^-6)$ e. $(3, ^-2)$ f. Parallel lines - no common solutions

    g. Same line - infinite number of common solutions.

5. a. Answers will vary. Slope has to be 1.

    b. Answers will vary.   c. $y - x = 3$, or an equivalent form

6. Answers will vary. Much teacher help and summary is needed here.

    a. For no common solutions, the two equations must have the same slope but different y-intercepts.
    b. For one common solution, the equations must have different slopes.
    c. For an infinite number of common solutions, one equation must be equivalent to the other.

. Susan laid her pencil on a
graph grid.

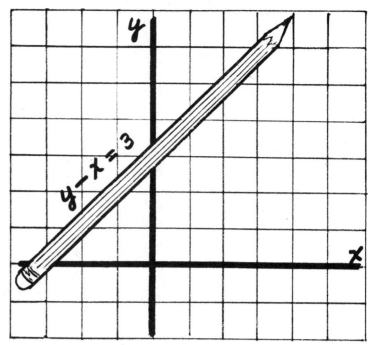

a. Use your pencil to show
a line parallel to
Susan's. What is the
equation of your line?

b. Use your pencil to show
a line that intersects
Susan's at (⁻1, 2).
What is the equation of
your line?

c. Use your pencil to show
a line that intersects
Susan's line in an
infinite number of points.
What is the equation of your line?

. Two straight lines are

parallel (no common solution) or

intersect in one point (one common solution) or

intersect in an infinite number of points (an infinite
number of common solutions).

a. Create a pair of equations that has no common solution.

b. Create a pair of equations that has exactly one
common solution (2,5).

c. Create a pair of equations that has an infinite number
of common solutions.

SOLVING SYSTEMS OF EQUATIONS - ANOTHER METHOD

$$x + y = 7$$
$$4x - y = 4$$

The graphs of each equation
are shown at the right. What do you
think the common solution is?
Does your guess fit the equations?

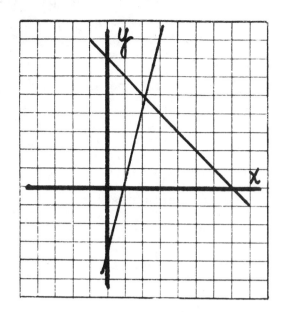

A graphing method is useful when an approximate solution is all
that is necessary. But suppose solutions involve fractions, and
an exact solution is required. In such cases a graphing method
may not be satisfactory.

These exercises will help you find a more efficient method for
solving a system of equations. A graphing procedure should
assist you in making certain discoveries.

1.

$$-x + y = 2$$
$$2x - y = 1$$

   a. Use graphing. Find the common solution to this pair of
       equations. Check your solution.

   b. Add the two equations together. You should get $x + 0 \cdot y = 3$
       or $x = 3$. Graph this new equation on the same grid.

   c. What have you noticed about the three graphs?

Solving Systems Of Equations - Another Method

Mathematics teaching objectives:

- Introduce adding as a method of solving systems of equations.
- Show that the graph of two equations and their sum have a common point of intersection.

Problem-solving skills pupils _might_ use:

- Study the solution process for clues.
- Make and use a graph.
- Make generalizations based upon data.

Materials needed:

- Graph paper

Background information to the teacher:

The graph shown at the beginning of the lesson demonstrates a limitation of the graphical method for solving a pair of equations. This is used as motivation for searching for another procedure. The exploration suggested in the lesson involves

- solving the pair of equations by graphing,
- adding the equations to obtain a third equation,
- graphing the third equation.

Pupils are expected to observe that the three graphs have the same intersection point. The graphs for the three equations are shown on page 252 as answers for problems 1 and 2. This characteristic is made more obvious if the graphs of the equations in problem 6 and their sum are graphed. The graphs of these three lines are shown in the chart accompanying the answer for problem 6. The dotted line is the graph of the third equation.

Comments and suggestions:

- Work with the class on problems 1-4a. Let pupils work independently or with a partner on problems 4b through 6. Let them struggle with problem 6, assuring them that the problem will be discussed at the next class meeting.

- Review adding expressions such as $\begin{array}{r} -x + y \\ \underline{2x - y} \end{array}$.

- Review the interpretation of $x = 3$ as the equation for a line perpendicular to the x-axis and $y = {}^-2$ as a line perpendicular to the y-axis.

- Elicit from the class a non-graphing procedure for problem 3. Expect pupils to explain how they found the value for _y_ after determining the value for _x_. Usually they use a substitution process similar to the problems they worked in the "I'm Thinking Of A Number" lesson.

- Give special attention to problem 5. It calls for a more efficient method for solving the introductory problem.

Answers:

See page 252, 253.

Solving Systems Of Equations - Another Method (cont.)

2.

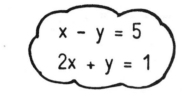

$$^-x + y = ^-6$$
$$x + y = 2$$

  a. Find the common solution by graphing.

  b. Add the two equations together. Graph the new equation on the same grid.

  c. What have you noticed about the three graphs?

3.

$$x - y = 5$$
$$2x + y = 1$$

  a. Add the two equations together.

  b. Without graphing the equations, try to use the new equation to find a solution to the original system.

  c. Graph the equations to check your method.

4. Use adding to find the common solution to each system of equations. Be sure to check each solution by substituting the values in the operations.

  a. $x + y = 3$ b. $2x + y = 0$ c. $5x + 3y = 10$

       $x - y = 5$ $^-2x + 3y = 8$ $2x - 3y = 4$

5. Find the common solution to the system of equations given at the first of the lesson.

6. Use your method for finding a common solution for these equations.

$$x + y = 4$$
$$3x - 2y = 7$$

If you have difficulty, graph them along with their sum. Explain why you had difficulties with your method.

PSM 82

Solving Systems Of Equations - Another Method

Answers:

1. a.

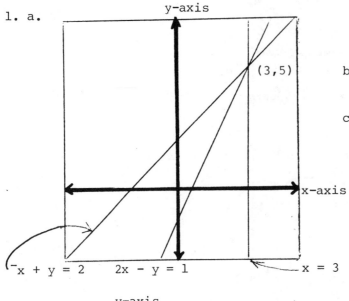

b. Pupils may need help in graphing x + 0 · y = 3 or x =

c. They all pass through (3,5).

2.a.

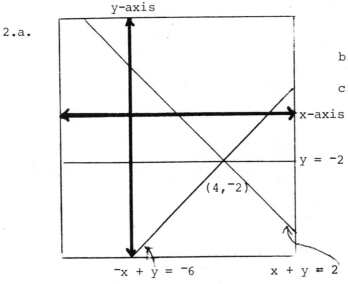

b. 0 · x + 2y = ⁻4 or 2y = ⁻4

c. They all pass through (4,⁻2)

3. a. 3x + 0 · y = 6 or 3x = 6

   b. Solve 3x = 6 to get x = 2. Substitute this value into one of the other equations to find ⁻3 for y.

4. a. (4,⁻1)

   b. (⁻1,2)

   c. (2,0)

5. $(2\frac{1}{5}, 4\frac{4}{5})$ or (2.2,4.8)

Solving Systems of Equations - Another Method

Answers:

*6.

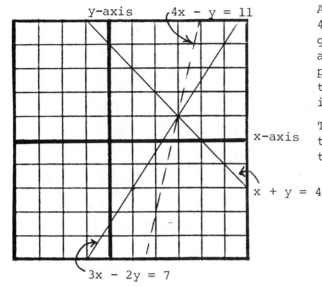

y-axis

4x - y = 11

x-axis

x + y = 4

3x - 2y = 7

Adding the two equations gives 4x - y = 11 as the sum. The graphing of this equation is an oblique line and therefore provides no clue as to either the x or y coordinates of the intersection point.

This problem leads nicely into the next step in the instructional process (see page 255).

(3, 1)

Problem 6 in the last lesson left pupils with the task of finding the solution
for the system

$$x + y = 4$$
$$3x - 2y = 7$$

They need to replace $x + y = 4$ by the equivalent equation $2x + 2y = 8$ before
adding the two equations to get a simplified third equation. The concept of
equivalent equations was introduced in the second lesson in this section (see
page 245), but no direct class assistance was provided for applying the concept
in the solution process. This move was left for pupils to make on their own.

1. Suggestions for the first follow-up lesson
   a. Elicit from the class accounts of their successes and/or frustrations
      with problem 6.
   b. Work with the class in making explicit statements on a procedure for
      solving a system requiring only one equation to be replaced by an
      equivalent one.
   c. Elicit from the class a procedure for solving a system using subtraction
      rather than addition, e.g.,

$$2x + 3y = 5$$
$$2x + y = {}^-1$$

   Work with the class in making explicit statements on a procedure for
   solving such systems.
   d. From a text assign exercises which provide practice with the procedures
      and at least one related word problem.
   e. Include as a final problem a system to be solved requiring both equations
      to be replaced by equivalent equations, e.g.,

$$3x + 2y = 17$$
$$4x - 3y = {}^-3$$

   Treat this problem in the same way problem 6 was in the last lesson.

2. Suggestions for the second follow-up lesson
   a. Elicit from the class accounts of their successes and frustrations with
      a system assigned in 1e.
   b. Work with the class on a procedure for solving a system requiring both
      equations to be replaced by equivalent equations.
   c. Assign from a text or other sources exercises which provide practice with
      the procedure. Choose exercises which require addition and subtraction.
      Also assign at least one related word problem.

d. Include as a final problem a system of equations written in the
y = mx + b form, e.g.   y = 3x + 5

$$y = 2x + 1$$

Treat this problem in the same way as problem 6 (see page 251). This
problem can be used in the next lesson to introduce the substitution
method for solving systems of equations.

3. Suggestions for the third follow-up lesson

a. Elicit from the class the methods they used for the problem assigned in
2d. They could use either the subtraction or substitution method.

b. Introduce the substitution method for solving systems including examples
with one equation in the y = mx + b form and another in the ax + by = c
form, e.g.
$$y = {}^-2x + 7$$
$$3x + y = 10$$

Some final comments:

Up to this point, pupils have not been required to use the substitution method
in solving systems with both equations written in the ax + by = c form. The
procedure involves the solution of either equation for x or y and then substitution into the other equation. This process is difficult for pupils and will
require at least two lessons. Possibly this can be deferred until the second
course in algebra.

Before the unit on systems is completed, several days should be spent on
solving word problems. Pupils also should be expected to write word problems
for a given system and to create word problems to be used in class assignments.

Algebra

X. CHALLENGES

X. CHALLENGES

The activities in the Getting Started section were very directed and pupils were encouraged to use (although not completely restricted to) one problem-solving skill at a time. The challenge problems in this section leave the choice of the problem-solving method up to the pupil. The intention is to allow for and encourage individual differences, creativity, and cooperation.

Let's look at how one challenge problem, "Good Friends," page 263, can be used in the classroom. As you read the example, notice how the teacher does not structure or direct the methods pupils use; but the teacher does have these important functions:

- to help pupils understand the problem.
- to listen if pupils want to discuss their strategies.
- to praise and encourage pupils in their attempts, successful or not.
- to facilitate discussion of the problem and sharing of the strategies.
- to give hints or ask questions, if necessary.
- to summarize or emphasize methods of solution after pupils have solved the problem.

As we look in Ms. Tengs' classroom, the page "Good Friends" has just been distributed. It is near the end of the math period--about 10 minutes left. Pupils are accustomed to seeing a different challenge each week and know they are expected to work on the problem on their own time, with some time given in class during the week to discuss their problems with others.

Ms. T: Here is the challenge problem of the week. I'll let you look at the problem, then we'll discuss it to be sure we all understand it. (Waits.) Charlie, you look like you have something to say.

Charlie: The first problem only asks us to solve the problem for five girls. Why don't we try to find the number of direct lines for 13 of them? That is what Mr. Ree is curious about! I'm going to work problem 2.

Ms. T: You can do that if you wish. Any other comments or suggestions? Marta?

Marta: Problem 1 is easier--I think I'll try it first.

Ms. T: Have you decided upon a strategy yet, Marta?

Marta: I think I'll name the girls Alice, Betty, Caron, Donna, and Ellen. Then I'll list and count all the direct lines to each home.

Ms. T: Good idea. Charlie, what do you plan to do?

Charlie: I think I'll draw a picture of the lines to 13 dots--the girls, and then I'll count the lines.

Ms. T: Drawing a picture also seems like a good idea. Keep working on the challenge and later in the week we'll form small groups to compare answers. I'll be interested in hearing about the strategies you used. (The period ends.)

See page 264 for suggestions for discussions throughout the week.

The above approach to challenge problems also gives opportunities for practicing the following problem-solving skills:

- State the problem in your own words.
- Clarify the problem through careful reading and by asking questions.
- Share data and the results with other interested persons.
- Listen to persons who have relevant knowledge and experiences to share.
- Study the solution process.
- Invent new problems by varying an old one.

THE TEACHER MUST BE AN ACTIVE, ENTHUSIASTIC SUPPORTER OF PROBLEM SOLVING.

Using the Activities

Twenty-four varied challenge problems are provided. Some teachers give them as a "challenge of the week" or as a Friday activity. On the day the challenge is given out, time should be spent on getting acquainted with the problem. On following days, a few minutes can be devoted to pupil progress reports. If there is little sign of progress, you can provide some direction by asking a key questi or suggesting a different strategy. At appropriate times, the activity can be summarized by a class discussion of strategies used and some problem extensions.

One of the skills listed on the back of nearly every challenge is "Make use of previous knowledge," specifying what algebraic background pupils should have "under their belts." This information should provide you with a clue as to when the challenge would best fit into your mathematics program.

Challenges can be used as a method of allowing for individual differences. However, the most value, if the development of problem-solving ability is the goal, is realized when pupils discuss their successful and unsuccessful strategie with others. At this time, they obtain clues as to how efficient their strategie were and in what ways the strategies might be improved.

<u>One</u> <u>Plan</u> <u>For</u> <u>Using</u> <u>A</u> <u>Challenge</u> (over a period of 1 or 2 weeks)

First day

. Give out the challenge. (Possibly near the end of the period.)

. Let pupils read written directions and possibly discuss with a classmate.

. Clarify any vocabulary which seems to be causing difficulty. Ask a few probing questions to see if they have enough understanding to get started.

. Remind them that during a later math class, time will be used to look at the problem again.

Later in the week

. Have pupils share their ideas.

. Identify the problem-solving skills suggested by these ideas.

. Conduct a brainstorming session if pupils do not seem to know how to get started.

. Suggest alternative strategies they might try.

. Give an extension to those pupils who have completed the challenge.

On a subsequent day

. Allow some class time for individuals (or small groups) to work on the challenge. Observe and encourage pupils in their attempts.

. Try a strategy along with the pupils (if pupils seem to have given up).

Last day

. Conduct a session where pupils can present the unsuccessful as well as the successful strategies they used.

. Possibly practice a problem-solving skill that is giving pupils difficulty, e.g., recording attempts, making a systematic list, or checking solutions.

Key problem-solving strategies that pupils have used in solving the problems are given in the comments for each problem. Your pupils might have additional ways of solving the problems.

A challenge problem for the teacher: Keep the quick problem solver from telling answers to classmates.

GOOD FRIENDS

Thirteen girls from Mr. Ree's algebra class are very good friends. They wish that each of them could have direct telephone lines to all the other girls. Mr. Ree is curious about the number of phone lines this would take.

1. Solve Mr. Ree's problem for five girls. The other eight girls are on vacation!

2. Solve the problem when all the girls are in town.

3. On a Saturday night each girl talks to all the others. Each girl finishes the call by saying "See ya Monday." How many times did "See ya Monday" get said?

4. Suppose fifteen girls wanted direct connections. How many phone lines would be needed?

5. All twenty-eight pupils in the class wanted direct lines to talk about their homework. How many lines are needed?

6. How many lines are needed for the principal to have a direct line to each of the 112 freshmen in the school?

7. How many lines are needed for all 337 pupils to have a direct connection to each of the other pupils?

Good Friends

Problem-solving skills pupils <u>might</u> use:

- Make use of previous knowledge--systematically recording data and using variables.

- Solve an easier but related problem.

- Use a drawing.

- Make a systematic list.

- Look for a pattern and generalize.

Materials needed:

- None

Comments and suggestions:

- The answer to the first problem, a simplification of the original problem, can be found rather quickly by making a drawing. This challenge lets pupils systematically increase the number of girls to be considered, make a systematic list, and then generalize.

- If no one seems to be using productive strategies, you may need to provide direction by asking key questions or making suggestions such as

  - Try making a drawing with a point representing each girl. How many direct telephone lines will it take for

    three girls?
    four girls?
    etc.

  - Find a way to record the results of your investigations and then search for patterns.

- During the summary session for the challenge, emphasize the importance of "working a simpler problem" as a useful skill to incorporate into problem-solving strategies.

- Pupils may recognize that some answers are triangular numbers. If they have previously developed the formula, $\frac{n}{2}(n - 1)$ the problems become application exercises. If triangular numbers have not been discussed in class, you may wish to help pupils use the following argument:

  There are <u>n</u> girls. To each girl's house there are n - 1 lines. Divide the product n(n - 1) by 2, since each line was counted twice.
  $\therefore$ the formula for the number of lines is $\frac{n(n - 1)}{2}$.

  Do not force this generalization if pupils are not ready for it.

Answers:

1. 10 lines
2. 78 lines
3. 156 times. Each girl says "See ya Monday" to the 12 other girls. Therefore, 13 x 12.
4. 105 lines
5. 378 lines
6. 112. Pupils may fail to see how this problem differs from nos. 4 and 5.
7. 56,616

A GAME OF CARDS

Kim has a special deck of cards. After shuffling the cards she turned six of them face up on the table.

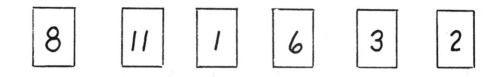

Kim explained the game this way:

> . Use each of the first five numbers in any order.
>
> . Use any operations you want.
>
> . Your answer must be the last card turned up (in this case, 2).

Here are two different solutions. Study them to make sure they're correct.

$$\frac{11 + 1}{3} - (8 - 6) = 2 \qquad \left(8 - \frac{1 + 11}{6}\right) \div 3 = 2$$

Kim dealt out five more hands. See how many different solutions you can find for each of them. Remember to express the results in correct mathematical form.

1. 3, 7, 1, 8, 13   _____ Answer is 6

2. 10, 15, 2, 10, 7   _____ Answer is 11

3. 4, 16, 7, 8, 9   _____ Answer is 1

4. 12, 7, 9, 7, 1   _____ Answer is 13

*5. 4, 25, 10, 22, 4   _____ Answer is 4

A Game Of Cards

Problem-solving skills pupils <u>might</u> use:

- Make use of previous knowledge--order of operations.
- Guess and check.
- Use appropriate mathematical notation.
- Work backwards.

Materials needed:

- Deck of 50 cards, two of each card labeled with a number from 1 to 25.

Comments and suggestions:

- Some of your pupils may remember this challenge from the PSM 8th-grade book. This activity is repeated here since it gives opportunities to construct appropriate arithmetic equations and to use the symbolism indicating order of operations.
- Rather than using this activity as challenges for individuals or small groups, you may wish to make a set of cards and actually do the activity with the entire class.
- You may wish to have pupils make a bulletin board display of different solutions.

Answers:

- Answers will vary.
- A possible solution for each problem is given.

1. $8 - \left[13 - (7 + 3) - 1\right] = 6$

2. $(10 \div 10)\left(\dfrac{15 + 7}{2}\right) = 11$ Note the use of the multiplication property of 1.

3. $\left[(8 \div \dfrac{16}{4}) + 7\right] \div 9 = 1$

4. $(7 - 7) \times 9 + 12 + 1 = 13$ Note the use of the multiplication property of 0.

*5. $(25 \div 10) + (22 \div 4) - 4 = 4$

Extensions:

- You might use numbers from ‾10 to 15 for the cards in the deck.
- Suggest that certain numbers be used as exponents. For example, for problem 5, the answer could be

$$\left(\dfrac{4}{4}\right)^{10} + (25 - 22) = 4$$

- Suggest the use of square roots.

$$\dfrac{10 + 22 + 4}{4 + \sqrt{25}} = 4$$

MULTIPLICATION REVERSALS

1. a. $\begin{array}{r} 64 \\ \times\,23 \\ \hline \end{array}$ b. $\begin{array}{r} 46 \\ \times\,32 \\ \hline \end{array}$ 2. a. $\begin{array}{r} 24 \\ \times\,84 \\ \hline \end{array}$ b. $\begin{array}{r} 42 \\ \times\,48 \\ \hline \end{array}$

3. a. $\begin{array}{r} 12 \\ \times\,63 \\ \hline \end{array}$ b. $\begin{array}{r} 21 \\ \times\,36 \\ \hline \end{array}$ 4. a. $\begin{array}{r} 93 \\ \times\,26 \\ \hline \end{array}$ b. $\begin{array}{r} 39 \\ \times\,62 \\ \hline \end{array}$

5. Write about your discovery._____

_____

6. What problem matches with $\begin{array}{r} 13 \\ \times\,93 \\ \hline \end{array}$?

7. Invent two more problems that use this discovery.

8. Use multiplication of two binomials to explain your discovery.

9. a. $\begin{array}{r} 253 \\ \times\,64 \\ \hline \end{array}$ b. $\begin{array}{r} 352 \\ \times\,46 \\ \hline \end{array}$ 10. a. $\begin{array}{r} 264 \\ \times\,84 \\ \hline \end{array}$ b. $\begin{array}{r} 462 \\ \times\,48 \\ \hline \end{array}$

11. Investigate problems 9 and 10 to discover when this will happen.

Multiplication Reversals

Problem-solving skills pupils <u>might</u> use:

. Make use of previous knowledge--multiplying binomials and solving equations.
. Look for patterns.
. Make generalizations.
. Guess and check.

Materials needed:

. Calculator (optional)

Comments and suggestions:

. Pupils may be more willing to try this challenge if they are allowed to use a calculator.
. The purpose of the challenge is not computation practice but the discovery of an interesting number pattern which can be demonstrated by using algebra. The significance of the pattern is "brought home" in problem 7 when pupils are asked to invent two problems which illustrate the number pattern discover:
. When the challenge is first introduced, provide enough class time for pupils to work problems 1 - 5. They should have an opportunity to explore the rest of the challenge on their own before sharing their results in small groups or with the total class.
. During the summary session, point out that frequently a problem which has been solved can be varied to make a new problem, e.g., problems 9 - 11.

Answers:

1.a,b. 1472   2.a,b. 2016   3.a,b. 756   4.a,b. 2418

5. At this point pupils probably will comment that the tens and units digits in the two digit factors have been reversed in part b. They also will notice that the products of both parts of each problem are the same. This will be true when the product of the tens digit in the two factors is equal to the product of the units digit. Pupils likely will not be aware of this latter statement until they have struggled with problem 7.

6.   31
    x 39

7. Answers will vary. (See answer to no. 5 above.)

8. Let $10a + b$ and $10c + d$ be the two-digit numbers. If
   $(10a + b)(10c + d) = (10b + a)(10d + c)$, then $ac = bd$.

$$
\begin{aligned}
(10a + b)(10c + d) &= (10b + a)(10d + c) \\
100ac + 10ad + 10bc + bd &= 100bd + 10bc + 10ad + ac \\
99ac &= 99bd \\
ac &= bd
\end{aligned}
$$

9.a,b. 16,192   10.a,b. 22,176

11. Impose the place value convention on the following notations.

    abc   cba   The products are the same if $ad = ce$
    x de   x ed   and $a + c = b$.

AN ALGEBRA CROSSNUMBER PUZZLE

1	2	3	4	5
6			7	
	8		9	10
11	12	13		
	14		15	

Each letter has a different whole number value.

a = ___, b = ___, c = ___, d = ___, e = ___, f = ___, g = ___

The heavy lines show where a number value stops.

	Across				Down		
1.	bd	11.	d^2	2.	f^2	9.	$(b + f)^2$
3.	a^5	13.	acf	3.	bg	10.	ce
6.	$13g$	14.	e^2	4.	$a^4 + d$	11.	abe
7.	$(ab)^2$	15.	a^3d	5.	$d^2 - b$	12.	ef
8.	$48b^2$			6.	ab^2	13.	ad

An Algebra Crossnumber Puzzle

Problem-solving skills pupils <u>might</u> use:

. Make use of previous knowledge--evaluating monomials and binomials with exponents.
. Guess and check.
. Eliminate possibilities.
. Record solution possibilities.

Materials needed:

. None

Comments and suggestions:

. Some class time will be needed to get the pupils into the problem. You might suggest 3-Across for a place to start. Pupils realize that <u>a</u> must be 2 since a^5 is a two-digit number. If <u>a</u> had any other whole number value it would not be a two-digit number. Let pupils work on their own before providing other clues.

. If after several minutes pupils have made no progress, you might give them some of the values; $a = 2$ and $b = 3$ will provide a good start.

. Additional clues you may need to provide later:
   - Focus attention on 7 across $(ab)^2$: If $a = 2$ and $(ab)^2$ must be a two-digit number, then b can be either 3 or 4. Try 3 first.
   - Focus attention on 6 down, ab^2: If $b = 3$, then $ab^2 = 18$.
   - Focus attention on 5 down, $d^2 - b$: If $b = 3$, then $d^2 - 3$ must equal a two-digit number ending with 6. This means that d must equal 7.

Answers:

1 2	2 1	3 3	4 2	5 4
6 1	4	3	7 3	6
8	4	3	9 2	10 4
11 4	12 9	13 1	2	0
8	14 6	4	15 5	6

One solution path:
3 across; $a = 2$
7 across suggests either 3 or 4 as a value for <u>b</u>. If 3 is chosen* the following is one sequence which will complete the puzzle.

6 down
5 down; $d = 7$
4 down
1 across
8 across
11 across
15 across
13 down
2 down; $f = 12$

9 down
13 across; $c = 5$
6 across; $g = 11$
3 down
11 down; $e = 8$
12 down, also 14
                     acrc

*If 4 is chosen instead, then $d^2 - 4$ must eq a two-digit number ending with 4. No whole number works.

Another possible solution path: could start as follows:
3 across; $a = 2$
6 across ⎫ $b = 3$
3 down ⎬ $g = 1$
2 down ⎭ $f = 12$
.
.
.

SEND MORE MONEY

Ross is in college. He's always writing his Dad for more money.
His last request was written as a puzzle problem.

```
  SEND
+ MORE
------
 MONEY
```

Dear Dad:

In this puzzle each different letter stands for a different digit. I know you can solve it Dad, because you're so clever. Furthermore, you're smart, ingenious, kind, benevolent, handsome, humorous... PLEASE SEND MONEY

Ross' Dad replied in a similar manner. The same rules apply.
However, the letters can have different values than they had in
the first puzzle.

```
  USE
+ LESS
------
 SONNY
```

Dear Son:

Your flattery will get you nowhere! However, if you can solve my puzzle, I might send you a few bucks. Incidentally, my puzzle has more than one solution. If you can find them all, I may send you a bit more.

EXTENSION

Ross' Dad collects foreign math books. He found a similar
puzzle in a book from Czechoslovakia.

```
  POSLI
+ IHNED
-------
 PENIZE
```

Translated, this means -

"Send immediately money."

<u>Send More Money</u>

Problem-solving skills pupils <u>might</u> use:
. Make use of previous knowledge--place value and a subtraction algorithm.
. Guess and check.
. Search for a place to start.
. List and eliminate possibilities.
. Search for other solutions.

Materials needed:
. None

Comments and suggestions:
. The second and third problems are considerably easier than the first. You might have pupils start with one of them.
. For USE + LESS = SONNY, some questions you might ask are:
  - What values could <u>S</u> have? <u>L</u>? <u>O</u>?
  - Give reasons why you selected these values.
  - Could the sum of <u>U</u> and <u>E</u> be less than 10? Why or why not?

. After these questions are dealt with let pupils work on their own. The next day pupils might give reasons about how they solved the puzzle.
. Here are some questions that might be asked to help pupils with the SEND + MORE = MONEY puzzle:
  - What values could <u>M</u> have? <u>O</u>? <u>S</u>?
  - Why can't S = 8?
  - Is N + R greater than 10? Why or why not?

Answers:

SEND		9567	M = 1	N + R > 10, since O = 0 and E + O ≠ E.
+ MORE	→	+ 1085	O = 0	Systematically let E be 2, 3, 4, 5, 6,
MONEY		10652	∴ S = 9	7, 8. Only E = 5 will work.

USE		715		814		517
+ LESS	→	+ 9511	or	+ 9411	or	+ 9711
SONNY		10226		10225		10228

POSLI		14568		19568
+ IHNED	→	+ 89302	or	+ 84302
PENIZE		103870		103870

WINDOW PANES

Sally works in a window factory. Her job is to program the computer. It tells the warehouse how many window panes to send to the assembly line.

Window panes come in three types:

   a. corner panes

   b. edge panes

   c. center panes

A 3 by 3 window is shown.
It uses 4 corner panes,
4 edge panes, and 1 center pane.

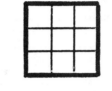

1. Make a table to show the number of panes of each type needed for these windows:

   a. 2 by 2   b. 3 by 3   c. 4 by 4   d. 5 by 5   e. 6 by 6

2. Study the table. How many panes of each type are needed for an _n_ by _n_ window?

3. Would the factory ever need window panes like these: ☐ or ☐ ? Why or why not?

4. The factory received a large order for rectangular windows, 2 by 3, 2 by 4, 2 by 5, etc. Help Sally decide how many panes of each type are needed. Include a 2 by _n_ window.

5. Investigate 3 by _n_ windows, 4 by _n_ windows, 5 by _n_ windows, and _m_ by _n_ windows.

Window Panes

Problem-solving skills pupils <u>might</u> use:

- . Make and use a drawing.
- . Make a systematic list.
- . Look for a pattern.
- . Make generalizations based upon data.

Materials needed:

- . None

Comments and suggestions:

- . Many patterns can be discovered in the data without having to go to the general case. For some pupils you may wish to stick with just the arithmetic examples, e.g., 2 by 2, 3 by 3, ... , 10 by 10.

- . A series of drawings is very effective for illustrating the patterns. Pupils easily see that in a 3 by n window the number of center panes is two less than the length <u>n</u>.

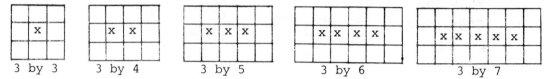

| 3 by 3 | 3 by 4 | 3 by 5 | 3 by 6 | 3 by 7 |

- . The drawings also show four corner panes are needed. The pattern for edge panes can be stated in several ways. Allow pupils the freedom to express the pattern in a way they understand. Some examples might be:

  - . 2 times the number of center panes plus one more on each end.
  - . the number of center panes plus one, then all multiplied by 2.
  - . total number of panes minus the number of center panes, then minus 4.

- . This problem is a 2-dimensional analogy to the famous "painted cube" problem found later in this section.

Answers:

1.

	2 by 2	3 by 3	4 by 4	5 by 5	6 by 6
corner	4	4	4	4	4
edge	0	4	8	12	16
center	0	1	4	9	16

2. n by n: corner panes, 4; edge panes, $4(n - 2)$; center panes, $(n - 2)^2$

3. ▢ would be needed for 1 by n windows $n \geq 2$.

   ▢ would be needed only for 1 by 1 windows.

4.

	2 by 3	2 by 4	2 by 5	2 by 6	2 by n
corner	4	4	4	4	4
edge	2	4	6	8	$2(n - 2)$
center	0	0	0	0	0

5.

	3 by n	4 by n	5 by n	m by n
corner	4	4	4	4
edge	$2(n - 2) + 2(1)$	$2(n - 2) + 2(2)$	$2(n - 2) + 2(3)$	$2(n - 2) + 2(m -$
center	$n - 2$	$2(n - 2)$	$3(n - 2)$	$(m - 2)(n -$

The notation for the edge panes indicates the number of panes along the top and the number on the side.

Problem 5 may need to be broken down into sub-problems and worked as problem 4 before going to the m x n case.

AN ODD HAPPENING

```
                1
              3   5
            7   9   11
         13   15   17   19
       21   23   25   27   29
```

1. Write the next two rows.

2. How many numbers in the
   a. 6th row?   c. 100th row?
   b. 10th row?   d. nth row?

3. Why does the 50th row not have a middle number?

4. What is the middle number in the
   a. 3rd row?   c. 7th row?   e. nth row?
   b. 5th row?   d. 25th row?

5. What is the difference in the first and last number of the
   a. 4th row?   c. 6th row?   e. nth row?
   b. 5th row?   d. 40th row?

6. What is the last number in the
   a. 5th row?   c. 40th row?
   b. 6th row?   d. nth row?

7. What is the sum of all the numbers in the
   a. 30th row?   b. nth row?

*8. What is the sum of all the numbers through the nth row?

An Odd Happening

Problem-solving skills pupils <u>might</u> use:

- . Make use of previous knowledge--polynomial expressions.
- . Look for patterns.
- . Make predictions based on generalizations.
- . Make and use a systematic list.

Materials needed:

- . None

Comments and suggestions:

- . This challenge involves many patterns which can be described with mathematics symbols and therefore may be worth investigating with the entire class. Possibly problems 1-4 could be done as class activities with the remaining to be done during a supervised study period.

- . Encourage pupils to make a systematic list or a table using the row number in order as the independent variable. The dependent variable would then be the second entry in the table.

- . Pupils likely will find it easy to produce the next pair in the table by examining the difference between successive entries. Such patterns do not lead directly to a formula for the nth term.

- . For pupils having difficulty with problem 6 you might suggest they compare the entry for the dependent variable with the square of the row number. Problem 8 is extremely difficult.

Answers:

1. Sixth row: 31 33 35 37 39 41
   Seventh row: 43 45 47 49 51 53 55

2. a. 6   b. 10   c. 100   d. n

3. The 50th row has an even number of elements. There are only middle numbers for those rows with an odd number of elements.

4. a. 9   b. 25   c. 49   d. 625   e. n^2

5. a. 6   b. 8   c. 10   d. 2(40 - 1) or 78   e. 2(n -

6. a. 29   b. 41   c. 1639   d. $n^2 + n - 1$

7. a. $30^3 = 27,000$ b. n^3 Have pupils look at simpler rows first.

*8. $\left[\dfrac{n(n + 1)}{2}\right]^2$ There is a close relationship between the required sum and the rule for generating triangular numbers.

Row	Cumulative Sum	
1	1	1^2
2	9	3^2
3	36	6^2
4	100	10^2
5	225	15^2
$\vdots$	$\vdots$	
n		$\left[\dfrac{n(n + 1)}{2}\right]^2$

THE TWELVE DAYS OF CHRISTMAS

The first day of Christmas
My true love sent to me
A partridge in a pear tree.

The second day of Christmas
My true love sent to me
Two turtle doves and
A partridge in a pear tree.

The third day of Christmas
My true love sent to me
Three French hens
Two turtle doves and
A partridge in a pear tree.

. . .
. . .
. . .

The twelfth day of Christmas
My true love sent to me
Twelve lords a leaping,
Eleven ladies dancing,
Ten pipers piping,
Nine drummers drumming,
Eights maids a milking,
Seven swans a swimming,
Six geese a laying,
Five gold rings.
Four colly birds,
Three French hens,
Two turtle doves, and
A partridge in a pear tree.

1. How many gifts were given on the third day of Christmas?

2.. Read the verse for the tenth day of Christmas.
   How many gifts were received that day?

3. "True love" gave many gifts of the same kind. Which kind of
   gift (or gifts) was given most often? How many?

4. How many total gifts were given during the twelve days of
   Christmas?

<u>The</u> <u>Twelve</u> <u>Days</u> <u>Of</u> <u>Christmas</u>

Problem-solving skills pupils <u>might</u> use:

. Make use of previous knowledge--triangular numbers.
. Make a systematic list.
. Look for patterns.

Materials needed:

. None

Comments and suggestions:

. As is usually the case with challenge activities there are different ways
the problem can be approached. Two different ways will be outlined along
with the answers. These, along with any others, should be discussed in
the culminating sessions.

. Pupils should be able to get into this exercise without much assistance.
However, it is likely they will want to discuss their solutions with other
persons in class.

Answers:

1. 6 gifts

2. The tenth day of Christmas $10 + 9 + 8 + 7 + \ldots + 2 + 1 = 55$ gifts
    My true love sent to me
    Ten pipers piping,
    Nine drummers drumming,
    Eight maids a milking,
    Seven swans a swimming,
    Six geese a laying,
    Five gold rings.
    Four colly birds,
    Three French hens,
    Two turtle doves, and
    A partridge in a pear tree

3. partridges $12 \cdot 1 = 12$ 42 geese or 42 swans
    turtle doves $11 \cdot 2 = 22$
    French hens $10 \cdot 3 = 30$
    colly birds $9 \cdot 4 = 36$
    gold rings $8 \cdot 5 = 40$
    geese $7 \cdot 6 = 42$
    swans $6 \cdot 7 = 42$
    maids $5 \cdot 8 = 40$
    drummers $4 \cdot 9 = 36$
    pipers $3 \cdot 10 = 30$
    ladies $2 \cdot 11 = 22$
    lords $1 \cdot 12 = 12$

4. 364 gifts.
    One way to find the total number of gifts is to add the column of
    figures in 3 above.
    Another way would be to find the sum of the first twelve triangle number
    $1 + 3 + 6 + 10 + 15 + 21 + \ldots + 66 + 78 = 364$ gifts.
    For your information, the general formula for the sum of the triangular
    numbers is $\frac{n}{6}(n + 2)(n + 1)$. The derivation of this formula is extremely
    difficult.

SOME SUMS

1. Place the digits 1, 2, 3, 4, 5, 6, and 7 in the circles so each line of connected circles has the same sum.

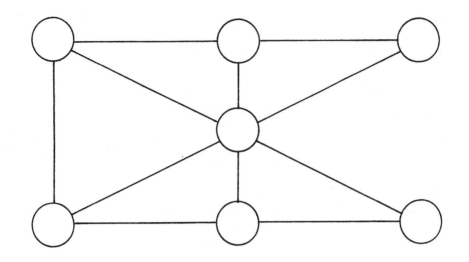

2. Repeat, using 2, 3, 4, 5, 6, 7, and 8.

3. Repeat, using 2, 4, 6, 8, 10, 12, and 14.

4. Repeat, using ⁻4, ⁻6, ⁻8, ⁻10, ⁻12, ⁻14, ⁻16.

5. Repeat, using 1, 3, 5, 7, 9, 11, and 13.

6. Find other numbers which could be used to solve the puzzle. Show where to place the numbers.

Some Sums

Problem-Solving skills pupils <u>might</u> use:

- . Make use of previous knowledge--addition with integers.
- . Guess and check.
- . Eliminate possibilities.
- . Make generalizations.

Materials needed:

- . None

Comments and suggestions:

- . You may wish to ask these questions when the problem is introduced:
  - How many different addition exercises are called for?
  - How many addends in each exercise?
- . You might ask these questions while pupils are working on the challenge:
  - In what way are the numbers used in each problem alike? How different?
- . During the culminating session, these questions or comments may stimulate further investigations:
  - Compare the solutions for problems 1, 2, 3, and 4.
  - In what way are the placements of numbers in the circles alike in each problem? How different?
  - Why can't the consecutive odd numbers (see problem 4) be used to solve the puzzle?

Answers:

1.

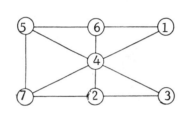

Notice that the numbers in two-addend sums are, on the average, larger than the numbers in the three-addend sums.

2.

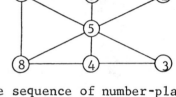

The sequence of number-placement here is more orderly than in problem 1.

3.

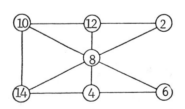

The sequence of number-placements is the same as in problem 1.

4.

The sequence of number-placement is the same as in problem 2.

5. Not possible. The problem has five 3-addend sums and one 2-addend sum. But the numbers given are all odds. Since 3 odds give an odd sum and 2 odds give an even sum, it is not possible to get all six sums to be the same.

6. The number-placement pattern for problems 1 and 3 is the same; call this a Type 1 placement. The number-placement pattern for problems 2 and 4 is the same; call this a Type 2 placement. Another solution set of numbers for either type can be generated by multiplying all numbers used in a solution by any integer.

THE PAINTED CUBE

1. Pat painted a 3 by 3 by 3 cube green on all six sides.

2. She then sliced the cube into 27 smaller cubes.

3. If Pat looked at each of the small cubes, how many would have paint on

   3 sides? ____

   2 sides? ____

   1 side? ____

   0 sides? ____

4. Place your results in the table below. Complete the table for other cubes.

	3 sides painted	2 sides painted	1 side painted	0 sides painted	total cubes
3 by 3 by 3					
4 by 4 by 4					
5 by 5 by 5					
6 by 6 by 6					
7 by 7 by 7					

The Painted Cube

Problem-solving skills pupils might use:

  . Make use of previous knowledge--polynomial expressions.
  . Visualize an object from its drawing.
  . Make and use a drawing or model.
  . Look for patterns.

Materials needed:

  . When working on the challenge, some pupils may prefer to use wooden
    cubes; three or four sets of 27 cubes should be enough.

  . You might color 27 cubes appropriately and let pupils try to assemble
    them into a 3 x 3 x 3 cube completely colored on all six sides.

Comments and suggestions:

  . Introduce the challenge by showing a physical model of 3 by 3 by 3
    cube made with unit cubes.

  . Some questions which might assist those pupils who are having difficulty
    finding the expressions for an n by n by n cube:

    - Where are the 3 "green-faced" cubes located? Two "green-faced" cubes?
      Single "green-faced" cubes?

    - What rule can be used to predict the number of cubes with no painted
      faces?

Answers:

3 & 4.

	3 sides painted	2 sides painted	1 side painted	0 sides painted	total cubes
3 by 3 by 3	8	12	6	1	27
4 by 4 by 4	8	24	24	8	64
5 by 5 by 5	8	36	54	27	125
6 by 6 by 6	8	48	96	64	216
7 by 7 by 7	8	$12(n-2)$	$6(n-2)^2$	$(n-2)^3$	n^3

Extension:

Refer back to the 3 by 3 by 3 cube (problem 1). What is the fewest number
of slices to separate the cube into 27 small cubes? A 4 by 4 by 4 cube
into 64 small cubes? A 5 by 5 by 5 cube into 125 small cubes?
An n by n by n cube into n^3 small cubes?

Answers: 3 by 3 by 3   6 slices
          4 by 4 by 4   9 slices
          5 by 5 by 5   12 slices
          n by n by n $3(n-1)$ slices

A YEAR-END SALE

1. A batting glove and a fielding glove together cost $44.00. The fielding glove cost 10 times as much as the batting glove. How much does each cost?

2. A ball and a bat together cost $11.00. The bat costs $10 more than the ball. How much does each cost?

3. Track City USA Sports Store is holding a Christmas sale.

    > Buy a football and a baseball for $ 18.00
    > Buy a baseball and a basketball for $ 22.00
    > Buy a basketball and a softball for $ 25.00
    > Buy a softball and a soccerball for $ 33.00
    > Buy a soccerball and a volleyball for $ 37.00
    > Buy a volleyball and a baseball for $ 17.00

    How much does each kind of ball cost?

    As you try to solve the problem, keep a careful, written record of what you tried.

A Year-End Sale

Problem-solving skills pupils <u>might</u> use:

. Guess and check.

. Make a systematic list.

. Eliminate possibilities.

Materials needed:

. None

Comments and suggestions:

. Give out this challenge with little or no comment except to mention the importance of keeping a careful record of step-by-step problem-solving efforts. This activity might be used as an informal assessment of the pupils' awareness of the processes they use to solve problems.

. Problems 1 and 2 are best solved by translating into an algebraic equation. Problem 2 often is missed by those using an intuitive approach. Such a person often says $10 and $1 which, of course, is only $9 more instead of $10 more.

. The third problem can be translated into 6 equations with 6 unknowns. However, the solution can be obtained much quicker by using an organized guess-and-check strategy. You may wish to point out that translating a problem into algebraic language is <u>NOT</u> the most efficient way to solve the problem.

Answers:

1. $40 and $4

2. $10.50 and $.50

3. football $13
    baseball $ 5
    basketball $17
    softball $ 8
    soccerball $25
    volleyball $12

THE TOWER OF BRAHMA

An ancient Hindu legend goes like this.

Brahma placed 64 disks of gold—each one a different size—in a stack so the largest disk was on the bottom. The temple priests were told to transfer the disks according to the following rules:

a. Only three stacks can be used.

b. Only one disk at a time can be moved.

c. No disk may be placed on top of a smaller disk.

d. Use the fewest moves possible.

The legend states that the world would vanish when the original stack of 64 disks was transferred to one of the other two stacks.

1. Guess the number of moves you think the priests would have to make. _____

2. Use the disks and playing board on the next page. Find the fewest number of moves to transfer a stack with 1 disk, 2 disks, 3 disks, etc. Complete the table.

3. How many years would it take for the priests to transfer a stack of 32 disks if they made one move per second?

Number in the stack	Fewest Moves
1	
2	
3	
4	
5	
.	
.	
.	
10	
.	
.	
.	
n	

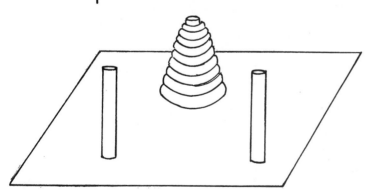

The Tower of Brahma

Problem-solving skills pupils <u>might</u> use:

. Make use of previous knowledge--expressions using exponents.

. Use a model.

. Make a systematic list.

. Look for patterns.

. Make predictions based on patterns.

Materials needed:

. The page of cut-outs accompanying the pupil page. A demonstration model would be nice to have. It could be made in a woodshop or could be purchased in a toy or craft store.

Comments and suggestions:

. Pupils need to be reminded that the determination of the fewest number of moves is not as easy as it seems. The difficulty begins in earnest when 5 disks are used.

. As pupils work at the puzzle they will find ways which can be used to decrease the number of moves. Encourage pupils to analyze and then verbalize their move strategies.

. As pupils discuss their move strategies with others, they most likely will come up with efficient strategies.

Answers:

Directions for a system for getting the minimum number of moves for any number of disks:

. Start with all disks on post A.

. Move the smallest piece in a <u>counter clockwise</u> direction to post B.

. Move the next smallest piece in a <u>clockwise</u> direction to post C.

. Move the smallest piece in a <u>counter clockwise</u> direction to post C.

. Move the third smallest piece to post B.

. Again, move the smallest piece in a <u>counter clockwise</u> direction. You need to remember the smallest piece moves every other time and always in the same direction. Also, when the second smallest piece moves, it always moves in a <u>clockwise</u> direction.

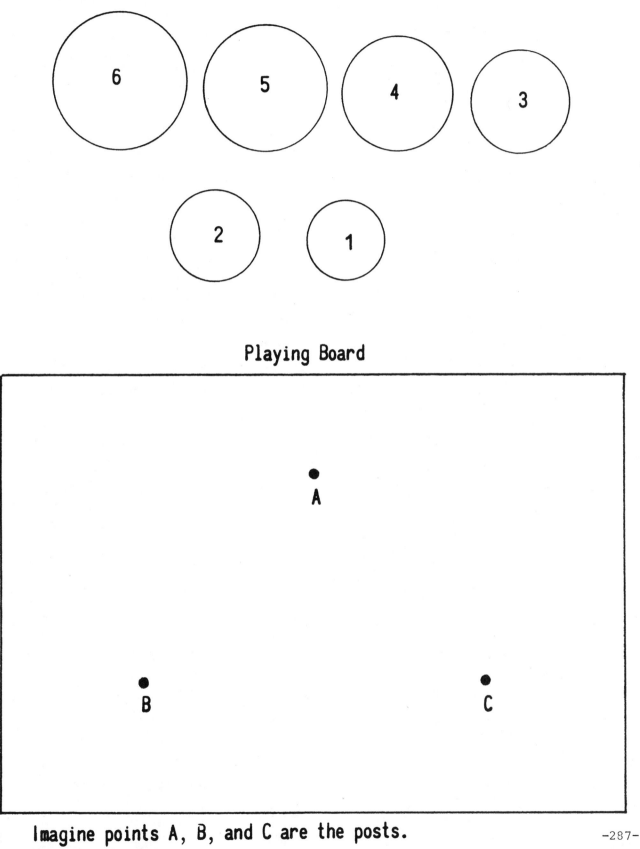

Playing Board

Imagine points A, B, and C are the posts.

The Tower of Brahma

Answers:

2.

A Number in the stack	B Fewest Moves
1	1
2	3
3	7
4	15
5	31
6	63
10	1023
n	$2^n - 1$

. It doesn't take long for pupils to observe that the next number in column B is always 2 times the last number plus one. For example,

$$31 = 15 \cdot 2 + 1.$$

They can understand this to be 15 moves to get 4 disks to post 1 move to get the fifth disk to post C, then 15 more moves to ge the remaining 4 disks to post C. course, another rule is necessary to get the expression for the fewest number of moves for n disks.

3. 4,294,967,296 seconds or a little more than 136 years.

Note: For the priests to move the 64 disks, it would take 18,446,744,073,709,551,615 seconds or over 584 billion years!

SAVINGS ACCOUNTS

Wanda and Roy each have $10 in the bank.

Every month Roy plans to deposit $1 in his account. Every month Wanda plans to deposit $4 in her separate account.

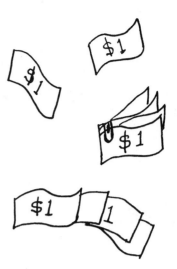

1. When will Wanda's account have

    a. twice as much as Roy's?
    b. three times as much?
    c. four times as much?

2. Suppose Roy and Wanda each started with $20 rather than $10. Each one still makes deposits as before. When will Wanda's account have

    a. twice as much as Roy's?
    b. three times as much?
    c. four times as much?

3. Suppose Wanda and Roy each started with $30. Again, each one makes deposits as before. When will Wanda's account be

    a. twice as much as Roy's?
    b. three times as much?
    c. four times as much?

4. Investigate similar problems when Roy deposits $1 per month but Wanda deposits $5 per month; or $6 per month; etc.

Savings Account

Problem-solving skills pupils _might_ use:

. Make use of previous knowledge--equation solving.
. Study a solution process.
. Make a systematic list.
. Make predictions based upon observed patterns.

Materials needed:

. None.

Comments and suggestions:

. This challenge can be solved several different ways including making a systematic list or solving equations. During the culminating discussion, these two methods could be contrasted. See comments in the answer sectio for other points that might be emphasized.

. When this challenge is introduced, call the pupils' attention to the way each problem and parts of each problem are "variations on a theme." As you will notice, they are expected to make their own variations as the last phase of this challenge. Some of the pupils' "invented" problems might be worth discussing with the total class.

Answers:

1.a. 5 months   b. 20 months   c. never

Systematic listing

	0	1	2	3	4	5	6	7	8	9	10	11	12	13	14	15
Roy	10	11	12	13	14	15	16	17	18	19	20	21	22	23	24	25
Wanda	10	14	18	22	26	30	34	38	42	46	50	54	58	62	66	70

17	18	19	20	21	22	...	n
27	28	29	30	31	32	...	$10 + n$
78	82	86	90	94	98	...	$10 + 4n$

Equation   1.a. $2(10 + n) = 10 + 4n$ 1.b. $3(10 + n) = 10 + 4n$
                  $20 + 2n = 10 + 4n$ $30 + 3n = 10 + 4n$
                  $n = 5$ months $n = 20$ months

1.c. $4(10 + n) = 10 + 4n$
       $40 + 4n = 10 + 4n$
               ???
       $40 = 10$

The systematic listing reveals that the amounts for both Wanda and Roy i crease as the months pass. However, it takes only 5 weeks for Wanda's savings to be twice that of Roy's but 20 weeks for hers to be 3 times as much. As pupils continue with the listing they will begin to realize th the amount for Wanda could never be 4 times as much as Roy's. This real zation becomes more convincing after examining the general term in the t No matter how large n becomes, $10 + 4n$ (Wanda's amount) can never be 4 t $10 + n$ (Roy's amount).

When the equation solution is examined it is immediately apparent that the equation has no solution.

2.a.   10 months   b. 40 months   c. never

3.a.   15 months   b. 60 months   c. never

Note: As the number of dollars each starts with changes in the various parts of problems 1, 2, and 3, the doubling time is always _half_ the initial amount and the tripling time is twice the initial amount. See the table for problem 1.

NUMBERS AND PATTERNS

Rectangular Numbers

1st   2nd   3rd   4th

Triangular Numbers

1st   2nd   3rd   4th

Square Numbers

1st   2nd   3rd   4th

Pentagonal Numbers

1st   2nd   3rd   4th

Complete the table.

	Rectangular Numbers	Triangular Numbers	Square Numbers	Pentagonal Numbers
1st	2			
2nd	6			
3rd	12			
4th				
5th				
100th				
Nth				

Numbers And Patterns

Problem-solving skills pupils <u>might</u> use:

. Make use of previous knowledge---operations with polynomials.
. Make and use tables and diagrams.
. Look for patterns.
. Make generalizations based upon data.

Materials needed:

. None

Comments and suggestions:

. Notice the concept of rectangular number is given a special meaning for this challenge. Usually a rectangular number is assumed to be any composite number. You may wish to comment about the "liberties" we have taken with this definition!

. Most pupils should be able to proceed with this challenge independently with little or no help until they get to the fifth row in the table. At this time, some work in small groups might be helpful.

. To check their N^{th} terms, pupils could substitute the number of each term. The expression is likely to be correct if, in each case, the corresponding rectangular number, triangular number, etc. is obtained.

. See answers for additional points that might be made when the challenge is discussed.

. Have pupils look for the patterns of the differences between successive entries in each column. Notice that for the rectangular numbers the successive differences are consecutive even numbers; for the triangle numbers, consecutive numbers; for the square numbers, consecutive odd numbers.

. Compare the sequence of numbers in each column. The triangular numbers can be obtained by dividing each rectangular number by 2. The pentagonal numbers can be obtained by adding, in a special way, a triangle number to a square number. Some examples: $1 + 4 = 5$, $3 + 9 = 12$, $6 + 12 = 22$, $10 + 25 = 35$. Many pupils will have difficulties getting the general term for pentagonal numbers by adding the expression for the $N - 1$ triangular number and the N square number. Be prepared to deal with this prior to the culminating session!

Answers:

	Rectangular Numbers	Triangular Numbers	Square Numbers	Pentagonal Numbers
1st	$4\binom{2}{6}$	$2\binom{1}{3}$	$3\binom{1}{4}$	1
2nd	$6\binom{}{12}$	$3\binom{}{6}$	$5\binom{}{9}$	5
3rd	$8\binom{12}{20}$	$4\binom{6}{10}$	$7\binom{9}{16}$	12
4th	$10\binom{20}{30}$	$5\binom{10}{15}$	$9\binom{16}{25}$	22
5th			25	35
100th	100(101)	100(101) ÷ 2	10,000	
Nth	$n(n + 1)$	$n(n + 1) \div 2$	n^2	$n^2 + \dfrac{(n - 1)}{2}$

*simplifies t

$$\frac{3n^2 - n}{2}$$

THE BOY AND THE DEVIL

u may recall the Boy and the Devil problem
rom the seventh grade. Now the boy is
der and wiser and the devil changed
ie problem slightly.

ee that bridge?" said the devil.
lust walk across and I will double
ur money. In fact, every time you
oss the bridge, I will double your
ney. But there is one condition.
u must pay me a little money after
ich crossing--this amount, though,
only $6."

is sounded reasonable to the boy. So
crossed the bridge, stopped to count
s money and, sure enough, the amount
d doubled. He paid the devil $6 and made another crossing.
ain his money doubled. He paid another $6 and crossed a third
me. Again his money doubled, but there was only $6 left. The
vil got the $6 and the boy had nothing!

How much money did the boy start with?

Suppose the problem remains the same except the boy gives up
only $4 after each crossing. Now how much money did the boy
start with?

Suppose the problem remains the same except that now the boy
gives up $30 after each crossing. Solve the problem by
using an equation. Let N be the amount of money the boy
started with.

Write a formula that can be used to solve any problem of this
kind.

   Let N be the amount the boy started with.
   Let X be the amount he must give up after each crossing.
   Check to see if your formula would solve problems 1-3.

<u>The</u> <u>Boy</u> <u>And</u> <u>The</u> <u>Devil</u>

Problem-solving skills pupils <u>might</u> use:

. Make use of previous knowledge--equation solving.
. Work backwards.
. Guess and check.
. Make a systematic record of solution paths.

Materials needed:

. None.

Comments and suggestions:

. To get the class started on the challenge you might suggest a guess and check approach. By "running through" a particular guess, the pupils get a better understanding of the structure of the problem. If pupils continue with the guess, check, and refine techniques they likely will want a less cumbersome procedure. The answer, which is an amount of money, is not an even-dollar amount.

. This challenge can be solved by working backwards and by solving equations. During the culminating discussion these methods could be contrasted and related. See comments in the answer section for other points that might be emphasized.

Answers:

1. $5.25

Three different approaches will be shown using the same vehicle--a carefully organized chart. Such a scheme does contrast and relate one approach with the others.

. Guess and check

	Before crossing	After crossing
1st trip	10	2(10) - 6 = 14
2nd trip	14	2(14) - 6 = 22
3rd trip	22	2(22)-- 6 = 38

Shows that a guess of $10 is too large.

	Before crossing	After crossing
1st trip	5	2(5) - 6 = 4
2nd trip	4	2(4) - 6 = 2
3rd trip	2	2(2) - 6 = ⁻2

Shows that a guess of $5 is too small.

Continued guessing is cumbersome but could be done.

. Work backwards

	Before crossing		After crossing		
1st trip	5.25	(10.50 ÷ 2)	10.50	- 6	= 4.50
2nd trip	4.50	(9 ÷ 2)	9	- 6	= 3
3rd trip	3	(6 ÷ 2)	6	- 6	= 0 ← Start here.

. Solve equations

	Before crossing	After crossing
1st trip	n	$2n - 6$
2nd trip	$2n - 6$	$2(2n - 6) - 6 = 4n - 18$
3rd trip	$4n - 18$	$2(4n - 18) - 6 = 0$ ←Now solve this equation.

wers (cont.)

2. $3.50

3. $2\left[2(2n - 30) - 30\right] - 30 = 0$

$\qquad \left[4n - 60 - 30\right] - 30 = 0$

$\qquad\qquad 8n - 180 - 30 = 0$

$\qquad\qquad\qquad 8n = 210$

$\qquad\qquad\qquad n = \$26.25$

Pupils may find it easier
to work through the chart
as shown above.

4. $2\left[2(2n - x) - x\right] - x = 0$

$\qquad\qquad 2(4n - 3x) - x = 0$

$\qquad\qquad\qquad 8n - 7x = 0$

$\qquad\qquad\qquad n = \dfrac{7}{8}x$

THE AGE OF DIOPHANTUS

Diophantus was a famous Greek mathematician who lived about 200 A.D. He has been called the "father of algebra" because of his contributions to that field. After his death a student composed this puzzle problem based upon his life.

His boyhood lasted for $\frac{1}{6}$ of his life.

His beard grew after $\frac{1}{12}$ more of his life had passed.

He married after $\frac{1}{7}$ more of his life had passed.

A son was born 5 years after his marriage.

The son lived half as many years as his father.

The father died 4 years after his son.

HOW OLD WAS DIOPHANTUS WHEN HE DIED?

Extension:

Create a similar puzzle in which a person's age can be determined.

<u>The</u> <u>Age</u> <u>Of</u> <u>Diophantus</u>

Problem-solving skills pupils <u>might</u> use:

- Make use of previous knowledge--solving equations with fractional coefficients.
- Guess and check.
- Translate a situation using algebraic symbols.

Materials needed:

- None

Comments and suggestions:

- The wording of the puzzle is "a bit involved" for most pupils. To those who seem lost, suggest that they check out a few guesses as a way for getting familiar with the puzzle. Guesses could lead to the solution especially if they realize that the answer must be divisible by 6, 7, 12, and 2.

Answer:

84 years old. $\frac{1}{6}n + \frac{1}{12}n + \frac{1}{7}n + 5 + \frac{1}{2}n + 4 = n$

SQUARES DO REPEAT

Get a calculator or a table of squares.

Study these problems. Observe the pattern in the answers.

$$27^2 + 68^2 = 5353 \qquad\qquad 47^2 + 66^2 = 6565$$

$$53^2 + 25^2 = 3434 \qquad\qquad 92^2 + 11^2 = 8585$$

Work these problems.

1. Which follow the pattern shown in the answers above?

   a. $18^2 + 79^2 =$ _____ c. $72^2 + 12^2 =$ _____

   b. $33^2 + 27^2 =$ _____ d. $61^2 + 4^2 =$ _____

2. Find the missing value.

   a. $42^2 +$ _____ $= 2020$ c. $27^2 +$ _____ $= 5353$

   b. $19^2 +$ _____ $= 8282$ d. $57^2 +$ _____ $= 7474$

3. What answer fits each of these first terms? Study the first term in the problems above. THINK SQUARES!!

   a. $36^2,$ _____ c. $77^2,$ _____

   b. $64^2,$ _____ d. $82^2,$ _____

4. Create other equations which fit the pattern.

   a. $(\underline{\quad})^2 + (\underline{\quad})^2 =$ _____ c. $(\underline{\quad})^2 + (\underline{\quad})^2 =$ _____

   b. $(\underline{\quad})^2 + (\underline{\quad})^2 =$ _____ d. $(\underline{\quad})^2 + (\underline{\quad})^2 =$ _____

*5. Does the pattern hold for any two-digit number? Try to find at least one exception. $(\underline{\quad})^2 + (\underline{\quad})^2 =$ _____

*6. Discover how the second term is determined. Study the first terms and THINK SUBTRACTION.

<u>Squares</u> <u>Do</u> <u>Repeat</u>

Problem-solving skills pupils <u>might</u> use:

. Guess and check.

. Look for a pattern.

. Work backwards.

Materials needed:

. Calculator or table of squares and square roots

Comments and suggestions:

. To make the computation aspect of this challenge easier, pupils will need a calculator or a table of squares and square roots.

. Problems 1 and 2 can be worked without knowing the pattern. Knowing the pattern allows completion of problems 3, 4, and 5. Some pupils will need help in seeing the pattern.

. For you only (or for pupils with extraordinary algebra skills), the proof goes like this: $(10a + b)^2 + (10b - a)^2 = 100a^2 + 20ab + b^2 + 100b^2 - 20ab + a^2 = 100a^2 + 100b^2 + a^2 + b^2 = 100(a^2 + b^2) + (a^2 + b^2)$.

Answers:

1. a. Yes - 6565   c. No - 5328

    b. Yes - 1818   d. Yes - 3737

2. a. 16^2 c. 68^2

    b. 89^2 d. 65^2

3. a. 4545   c. 9898

    b. 5252   d. 6868

4. Answers will vary.

*5. No, only those such that the sum of the squares $\leq$ 99.

*6. If the two-digit number is 10a + b, the second term will be 10b - a.

PRUNES FOR DESSERT

1. Three brothers--Jim, Mike, and Eric--after finishing a meal in a restaurant, ordered a bowl of stewed prunes. While waiting for the prunes to be served, all three fell asleep. After a while, Jim woke up and found the prunes on the table. He ate his equal share and went back to sleep. Then Mike awoke, ate what he thought was his equal share, and fell asleep again. Eric woke up, ate what he thought was his equal share of the remaining prunes, and went back to sleep.

   A little while later, all three brothers woke up and discovered that eight prunes were left in the bowl.

   How many prunes were in the bowl originally? (No fractions allowed on prune problems.)

2. The next day, Steve joined the group for dinner. Again they ordered prunes and fell asleep. As before, they awoke one at a time and each ate $\frac{1}{4}$ of what was left in the bowl. When they awakened they discovered 81 prunes remained.

   How many prunes were in the bowl originally?

EXTENSION

First, fill in the blank and then answer the question.

The next day, Ted joined the group. The same thing happened as before. This time, however, they found _____ prunes after they all awakened.

How many prunes were in the bowl originally?

Prunes For Dessert

Problem-solving skills pupils <u>might</u> use:

. Make use of previous knowledge--equation solving involving fraction coefficients.
. Work backwards.
. Look for clues by studying a previously-solved problem.
. Look for patterns.

Materials needed:

. None.

Comments and suggestions:

. If additional clues are needed to get pupils into the challenge, you might try using a chart as in The Boy And The Devil challenge. Have pupils "run through" particular guesses.

    Pupils will quickly notice random guessing usually gets them a fractional answer--a "no-no" for this problem. A chart follows along with a guess of 30 prunes.

	Before nap	After nap
Jim	30	$30 - \frac{1}{3}(30) = 20$
Mike	20	$20 - \frac{1}{3}(20) = 13\frac{1}{3}$
Eric		

A guess of 30 leads to a fractional answer of $13\frac{1}{3}$.

    When pupils encounter such difficulties they may suggest working backwards through the table or selecting guesses which have 3 factors of three.

Answers:

    The equations for solving the sub-problems are given below along with the answers.

1. 27

$$\frac{2}{3}\left(\frac{2}{3}\left(\frac{2}{3}n\right)\right) = 8$$

$$\frac{8}{27}n = 8$$

$$n = 27$$

2. 256

$$\frac{3}{4}\left(\frac{3}{4}\left(\frac{3}{4}\left(\frac{3}{4}n\right)\right)\right) = 81$$

$$\frac{81}{256}n = 81$$

$$n = 256$$

Extension:

$$\frac{4}{5}\left(\frac{4}{5}\left(\frac{4}{5}\left(\frac{4}{5}\left(\frac{4}{5}n\right)\right)\right)\right) = x$$

$$\frac{1024}{3125}n = x$$

    Since no fractional prune is allowed, the least value for (n,x) is (3125,1024). Other values are (6250,2048); (9375,3072); etc.

WHAT WOULD HAPPEN IF ...

1. Draw the graph of

    $$|x| + |y| = 6.$$

2. What would happen if we changed the equation to

    $$|x| - |y| = 6?$$

    First make a guess. Then check it out by graphing.

3. What would happen if we changed the equation to

    $$|y| - |x| = 6?$$

    Make a guess. Then check it out by graphing.

4. What would happen if we changed the equation to

    $$|x + y| = 6?$$

    Make a guess. Then check it out by graphing.

5. Invent some of your own "What would happen ifs." Keep a
    record of your results.

<u>What Would Happen If ...</u>

Problem-solving skills pupils <u>might</u> use:

   . Make use of previous knowledge--absolute value and graphing.
   . Make and use a table (systematic listing).
   . Make predictions based upon previous experiences.
   . Invent new problems by varying ones already solved.

Materials needed:

   . Graph paper (see page 306)

Comments and suggestions:

   . This challenge can be used to build skill in using the absolute value
     symbolism. Probably the best time to introduce the challenge is after
     the class has had the first problem $|X| + |Y| = 6$ as a regular assignment.
     During this time pupils should have dealt with the equivalent equation
     $|Y| = 6 - |X|$ and the impossibility of replacing X by any number greater
     than 6.

   . In making a table of values for these equation, given values for X may have
     0, 1, or 2 values for Y. Pupils may not have encountered this possibility
     before.

   . As you have opportunities to discuss this challenge with individuals,
     question them on the success they have in guessing what happens. Their
     success should improve as they make guesses and move from problem to problem.

Answers:

1.   2.

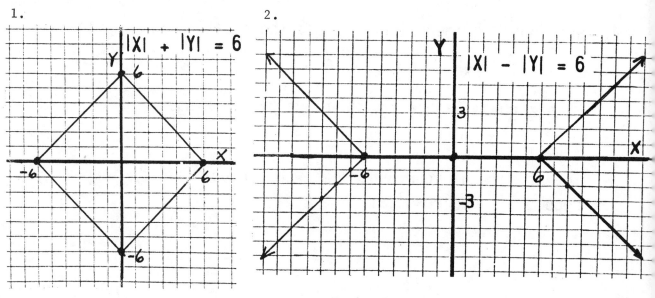

Notice how the change in operation
from addition to subtraction turns
the graph "inside out."

Answers: (cont.)

3.   4.

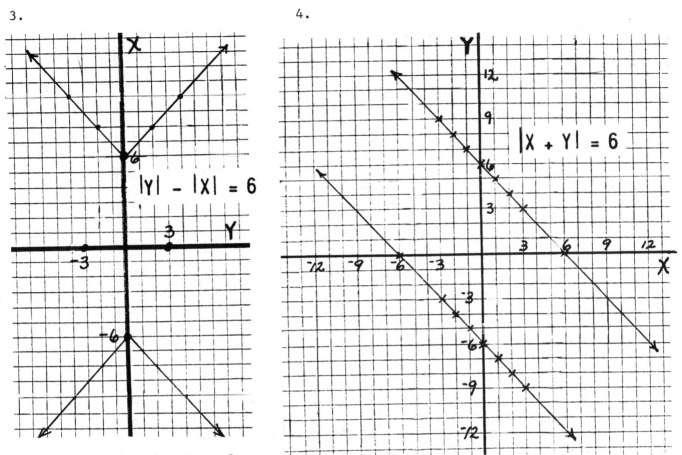

Notice how the commuting of
X and Y rotates the graph
through 90°.

Notice how the change in absolute value
designations eliminates two sides of the
square shown in problem 1.

What do you think will happen when a similar
change is made in problem 2?

5. Other problems pupils might suggest are $|X - Y| = 6$; $|X| - Y = 6$;
    $X + |Y| = 6$; etc.

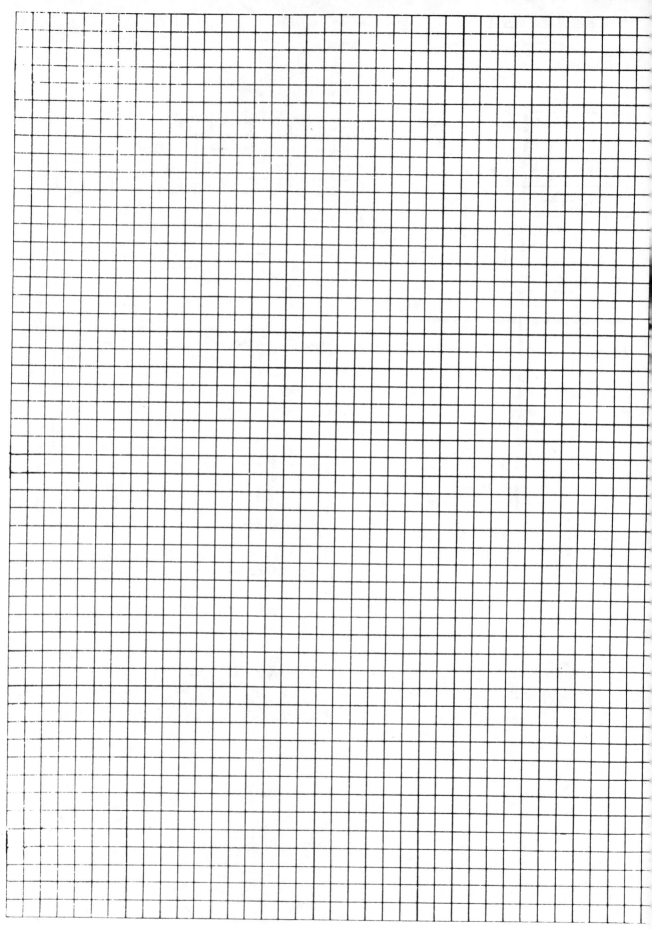

UNIT FRACTIONS

Unit fractions always have 1 for a numerator.

1. Work these problems. Look for patterns in the denominators.

   a. $\dfrac{1}{6} + \dfrac{1}{3} = \dfrac{1}{\boxed{}}$

   b. $\dfrac{1}{30} + \dfrac{1}{6} = \dfrac{1}{\boxed{}}$

   c. $\dfrac{1}{20} + \dfrac{1}{5} = \dfrac{1}{\boxed{}}$

2. Use the above pattern to predict these missing denominators. Check.

   a. $\dfrac{1}{\boxed{}} + \dfrac{1}{7} = \dfrac{1}{6}$

   b. $\dfrac{1}{12} + \dfrac{1}{\boxed{}} = \dfrac{1}{3}$

   c. $\dfrac{1}{90} + \dfrac{1}{10} = \dfrac{1}{\boxed{}}$

3. Write three more problems having the same pattern as above,

   a. $\dfrac{1}{\rule{1em}{0.4pt}} + \dfrac{1}{\rule{1em}{0.4pt}} = \dfrac{1}{\rule{1em}{0.4pt}}$

   b. $\dfrac{1}{\rule{1em}{0.4pt}} + \dfrac{1}{\rule{1em}{0.4pt}} = \dfrac{1}{\rule{1em}{0.4pt}}$

   c. $\dfrac{1}{\rule{1em}{0.4pt}} + \dfrac{1}{\rule{1em}{0.4pt}} = \dfrac{1}{\rule{1em}{0.4pt}}$

4. Write an algebraic statement to describe the pattern.

   $$\dfrac{1}{\rule{3em}{0.4pt}} + \dfrac{1}{\rule{3em}{0.4pt}} = \dfrac{1}{a}$$

5. Work these problems. Look for patterns in the denominators.

   a. $\dfrac{1}{12} + \dfrac{1}{6} + \dfrac{1}{4} = \dfrac{1}{\boxed{}}$

   b. $\dfrac{1}{40} + \dfrac{1}{10} + \dfrac{1}{8} = \dfrac{1}{\boxed{}}$

6. Use the above pattern to predict these missing denominators. Check.

   a. $\dfrac{1}{\rule{1em}{0.4pt}} + \dfrac{1}{\rule{1em}{0.4pt}} + \dfrac{1}{\rule{1em}{0.4pt}} = \dfrac{1}{5}$

   b. $\dfrac{1}{\rule{1em}{0.4pt}} + \dfrac{1}{\rule{1em}{0.4pt}} + \dfrac{1}{\rule{1em}{0.4pt}} = \dfrac{1}{3}$

7. Write an algebraic statement to describe the pattern.

   $$\dfrac{1}{\rule{3em}{0.4pt}} + \dfrac{1}{\rule{3em}{0.4pt}} + \dfrac{1}{\rule{3em}{0.4pt}} = \dfrac{1}{a}$$

<u>Unit Fractions</u>

Problem-solving skills pupils <u>might</u> use:

- Look for a pattern.
- Make an algebraic explanation.

Materials needed:

- None

Comments and suggestions:

- Pupils need addition of arithmetic fractions as a prerequisite for this activity.
- Pupils are asked to just <u>write</u> an algebraic statement to <u>describe</u> the patterns. Better pupils should be asked to show that the statement involving the algebraic fractions does give the indicated sum.
- Many patterns may be found. Accept those for which a pupil has a reasonable explanation.

Answers:

1. a. $\dfrac{1}{2}$ b. $\dfrac{1}{5}$ c. $\dfrac{1}{4}$

2. a. $\dfrac{1}{42}$ b. $\dfrac{1}{4}$ c. $\dfrac{1}{9}$

3. Answers will vary.

4. $\dfrac{1}{a(a+1)} + \dfrac{1}{a+1} = \dfrac{1}{a}$

5. a. $\dfrac{1}{2}$ b. $\dfrac{1}{4}$

6. a. $\dfrac{1}{60} + \dfrac{1}{12} + \dfrac{1}{10} = \dfrac{1}{5}$ b. $\dfrac{1}{24} + \dfrac{1}{8} + \dfrac{1}{6} = \dfrac{1}{3}$

7. $\dfrac{1}{a(2a+2)} + \dfrac{1}{2a+2} + \dfrac{1}{2a} = \dfrac{1}{a}$

THE TORTOISE AND THE HARE

The Hare challenged the Tortoise to the best two out of three races. In each case, the race was to be 100 metres.

Race 1

The Tortoise left the starting line, "sprinted" at the rate of 4 m per minute. Twenty-five minutes later the Hare left the starting line. How fast did the Hare have to run in order to overtake the Tortoise?

Who won the first race?

Race 2

The Hare left 8 minutes after the Tortoise. The Hare ran at the rate of 20 m per minute. The Tortoise still sprinted at 4 m per minute.

Draw a graph that shows the progress of the race. Use the same grid for both the Tortoise and the Hare. The horizontal axis should show the time; the vertical axis should show the distance.

Who won the second race?

Race 3

The Hare left 5 minutes after the Tortoise. After the Hare ran for 3 minutes it stopped for a 15 minute rest and then resumed the race. The Tortoise still sprinted at 4 m per minute and the Hare ran at 20 m per minute.

Make another graph to show the progress of each.

Who won the third race?

The Tortoise And The Hare

Problem-solving skills pupils <u>might</u> use:

. Make use of previous knowledge--graphing and selecting appropriate scales.
. Make and use graphs to solve a problem.
. Clarify the problem through careful reading.

Materials needed:

. Graph paper (see page 306)

Comments and suggestions:

. Pupils usually do not get practice solving a problem graphically by moving from the verbal statement of the problem directly to the plotting of necessary point to make the graph. They probably will have difficulty. Let them struggle for a while working individually and in small groups before you offer assistance.

. Decisions need to be made concerning the scales for the horizontal and vertical axes. The graph becomes almost unmanageable if the scale on the vertical axis is the same as on the horizontal. After a while you may suggest they try different scales for each axis.

Answers:

<u>Race 1</u>. The tortoise is at the finish line when the hare starts.
<u>Race 2</u>. The hare wins.   <u>Race 3</u>. The hare and the tortoise tie!

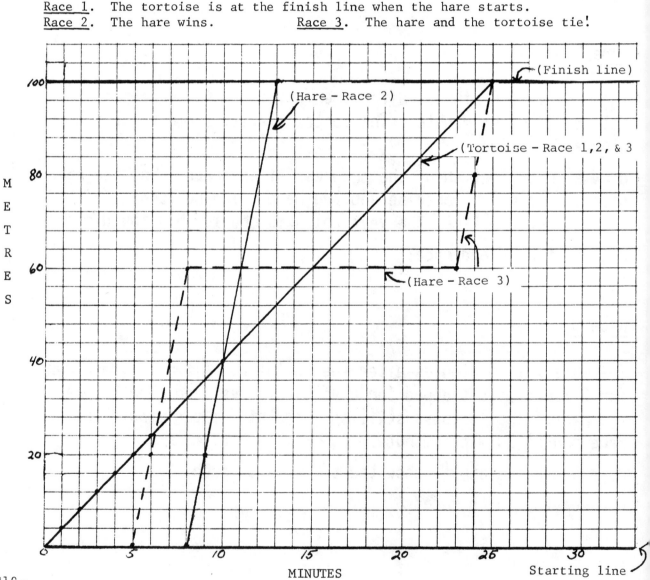

HOW MANY CUBES?

This cube is 3 by 3 by 3.

1. How many of these cubes can be found in the 3 by 3 by 3 cube?

   a. 1 by 1 by 1 _____
   b. 2 by 2 by 2 _____
   c. 3 by 3 by 3 _____

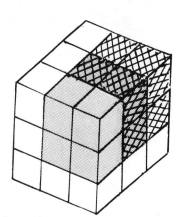

One 2 by 2 by 2 cube
is shaded. ▨
Another 2 by 2 by 2 cube is shaded. ▩

2. Complete this table. Look for patterns.

Cube Size	Number of Cubes					
	1 by 1 by 1	2 by 2 by 2	3 by 3 by 3	4 by 4 by 4	5 by 5 by 5	n by n by n
by 1 by 1						
by 2 by 2						
by 3 by 3						
by 4 by 4						
by 5 by 5						
by n by n						

3. How many cubes, each different, could be cut from an 8 by 8 by 8 cube.

How <u>Many</u> <u>Cubes</u>?

Problem-solving skills pupils <u>might</u> use:

. Make use of previous knowledge--polynomial expressions and volume concept

. Visualize an object from its drawing.

. Make and use a drawing or model.

. Make and use a table.

. Look for patterns.

Materials needed:

. Some pupils may prefer to work on the challenge using wooden cubes. One set of 125 cubes should be enough.

Comments and suggestions:

. Have pupils read over parts 1 and 2 of this challenge and indicate in which way this challenge is different from the painted-cube challenge.

. Some time should be spent on part 2 explaining the diagram showing two shaded cubes. In counting the different cubes, we must consider "over-lapping" cubes.

. Let pupils proceed on their own after working on parts 1 and 2 with the total class.

Answers:

1. a. 27   b. 8   c. 1

2.

	Number of Cubes					
Cube Size	1 by 1 by 1	2 by 2 by 2	3 by 3 by 3	4 by 4 by 4	5 by 5 by 5	n by n by n
1 by 1 by 1	1					
2 by 2 by 2	8	1				
3 by 3 by 3	27	8	1			
4 by 4 by 4	64	27	8	1		
5 by 5 by 5	125	64	27	8	1	
n by n by n	n^3	$(n-1)^3$	$(n-2)^3$	$(n-3)^3$	$n(-4)^3$	1

3. $8^3 + 7^3 + 6^3 + 5^3 + 4^3 + 3^3 + 2^3 + 1^3 = 1296$

NUMBERS IN BOXES

numbers inside the dotted boxes can be found by looking for
terns. Discover the pattern and then finish problems c and d.

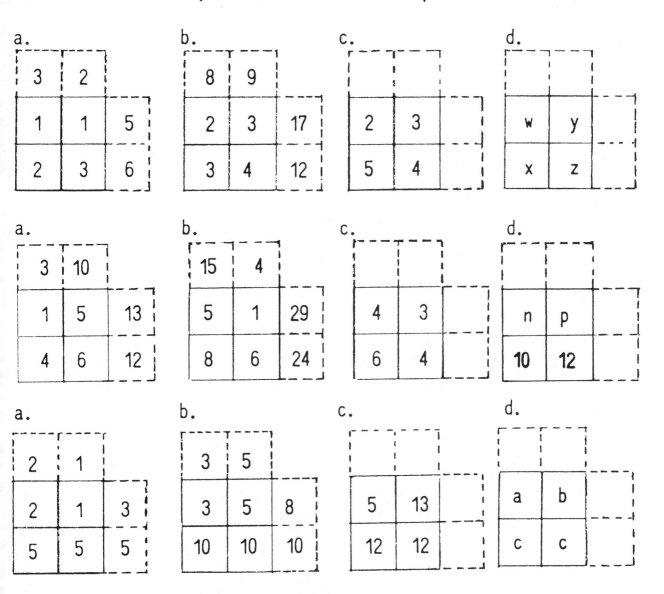

a.

3	2	
1	1	5
2	3	6

b.

8	9	
2	3	17
3	4	12

c.

2	3	
5	4	

d.

w	y	
x	z	

a.

3	10	
1	5	13
4	6	12

b.

15	4	
5	1	29
8	6	24

c.

4	3	
6	4	

d.

n	p	
10	12	

a.

2	1	
2	1	3
5	5	5

b.

3	5	
3	5	8
10	10	10

c.

5	13	
12	12	

d.

a	b	
c	c	

Write your own variables in boxes. Do _a_ like problem 1,
_b_ like problem 2, and _c_ like problem 3. Be sure to fill in the
dotted squares.

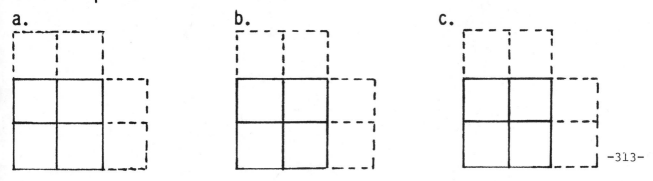

a.

b.

c.

Numbers In Boxes

Problem-solving skills pupils <u>might</u> use:

- Look for a pattern.
- Make an algebraic explanation.
- Create new problems by varying a given one.

Materials needed:

- None

Comments and suggestions:

- The activity is a disguised method of adding fractions. Most pupils will not recognize this and will solve the problems by seeing patterns.
- Several patterns might be found. Accept all those for which a pupil has a reasonable explanation.
- Number 4 is intended to be solved as algebraic fractions but pupils may find it easier to create an arithmetic example.

Answers:

1.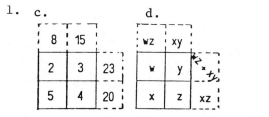

2.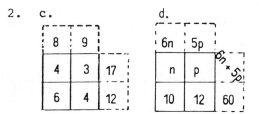

3.

4. Answers will vary. A sample is given.

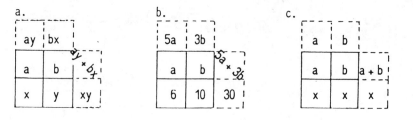

FREE THROWS

1. "Dead Eye" Joe normally is the best free throw shooter on the team. Yet, so far this season, he has made only 9 out of 20. How many consecutive free throws must "Dead Eye" make to raise his record to 75%?

2. "Air Ball" Bud normally is the worst free throw shooter on the team. Yet, so far this season, his record is fantastic--17 out of 21. His slump is sure to begin. How many consecutive free throws will "Air Ball" have to miss for him to lower his record to 25%? (Of course, Coach Wynn M. Awle hopes this will never occur.)

3. Create a problem about "Sure Shot" Sue who normally makes 50% of her free throws but now is making just 20%.

4. Create a problem about "Slam Dunk" Sam who normally makes 50% of his free throws but now is making 80%.

5. "Leaper" LeAnn has made just 3 of 17 free throws. She wants to get her percent up to 90%. If she makes 5 out of 5 shots in each of her remaining games, how many games will it take to pass the 90% mark?

Problem-solving skills pupils <u>might</u> use:

- . Make use of previous knowledge--fractional equations and percent.
- . Guess and check.
- . Make a systematic list.
- . Describe situations using algebraic symbolism.

Materials needed:

- . None

Comments and suggestions:

- . When the challenge is introduced, have pupils guess the number of consecutive "hits" that would be needed. As pupils check, they will see that they need to add the "answer" to both numerator and denominator before converting the ratio to percent. After a few guesses, let the pupils work on their own.

- . Culminate the activity by eliciting from pupils any equations they used to solve the problems.

Answers:

1. "Dead Eye" Joe must make 24 consecutive baskets.

   Equation: $\dfrac{9 + x}{20 + x} = \dfrac{75}{100}$ $x = 24$

2. "Air Ball" Bud must make 47 consecutive misses.

   Equation: $\dfrac{17}{21 + x} = \dfrac{25}{100}$ $x = 47$

3. Answers will vary.

4. Answers will vary.

5. "Leaper" LeAnn needs 25 games to get to the 90% mark.

   Equation: $\dfrac{3 + 5x}{17 + 5x} > \dfrac{90}{100}$, $x > 25$